XFORD SCIENCE RESEARCH PAPERS

Solvable Models in Algebraic Statistical Mechanics

D. A. Dubin

SOLVABLE MODELS IN ALGEBRAIC STATISTICAL MECHANICS

BY

D. A. DUBIN

CLARENDON PRESS • OXFORD

1974

Oxford University Press, Ely House, London W. 1

GLASGOW NEW YORK TORONTO MELBOURNE WELLINGTON
CAPE TOWN IBADAN NAIROBI DAR ES SALAAM LUSAKA ADDIS ABABA
DELHI BOMBAY CALCUTTA MADRAS KARACHI LAHORE DACCA
KUALA LUMPUR SINGAPORE HONG KONG TOKYO

ISBN 0 19 853341 1

Printed in Great Britain by
J. W. Arrowsmith Ltd., Bristol

Preface

THE PURPOSE of this book is to gather together the widely scattered material concerning some exactly solvable models of statistical-mechanical systems with an infinite number of degrees of freedom. By a process of selection and arrangement we hope that we have presented this material in a coherent and uniform setting.

It is assumed, but it is not necessary, that the reader is acquainted with the physical motivation behind the models discussed, for our analysis of such matters is necessarily brief (Boltzmann [1, 2]; Born [1]; Ehrenfest and Ehrenfest [1]; Gibbs [1]; Guggenheim [1]; Huang [1]; Landau and Lifschitz [1]; London [1]; Mattis [1]; Schrödinger [1]).

We have also assumed a concurrent—or prior—knowledge of the pertinent material in the textbooks of Emch [1] and Ruelle [1]. Another good reference is the text of Eckmann and Guenin [1]. Any supplementary mathematics required of the reader will be found in the various textbooks referred to in the body of this text.

The readers we had in mind during the writing were, primarily, graduate students embarking on research projects in statistical mechanics. In addition, it is hoped that this book will be of use to the general community of research workers in statistical mechanics and applied mathematics.

With this in mind, in spite of the rather technical nature of the material, very few detailed proofs are presented. Most statements and theorems are proven 'by reference', generally to a journal article. For it is not expected that the reader we had in mind will wish to know all such details, particularly at first reading. If we have managed to convey the general flavour of the subject and enough background to enable the reader to find his way through the literature, then we shall have achieved our aim.

A book such as this is almost surely plagued with errors; but I am not the first author to be faced with this problem. Concerning his dictionary (Johnson, S. (1755). *A dictionary of the English language*), Samuel Johnson defended his authorship thus: '. . . a few wild blunders and risible absurdities, from which no work of such multiplicity was ever free, may for a time furnish folly with laughter . . . What is obvious is not always known, and what is known is not always present. In this work, when it shall be found that much is omitted, let it not be forgotten that much is likewise performed.'

Thornborough,
April, 1973

D.A.D.

Acknowledgements

I would like to express my gratitude to G. G. Emch, J. T. Lewis, R. F. Streater, M. Winnink, and especially G. L. Sewell for some conversations about algebraic statistical mechanics which I had with them at various times. I am particularly indebted to E. B. Davies for undertaking the burdensome task of critically reading the manuscript and for making a number of helpful suggestions.

G. G. Emch and E. Montroll arranged for me to spend a semester at the University of Rochester, where almost all of the final draft was written; their warm hospitality is much appreciated.

The first three chapters of the manuscript were typed by Clare Belchamber; the remainder was typed at the Mathematics Faculty of the Open University. It is scant acknowledgement to thank Larrie Ridley-Thomson, Gladys Howe, Barbara Keogh, and Frances Thomas for their assistance.

Finally, I wish to thank the Clarendon Press for their patience and understanding concerning the manuscript.

Contents

1. THE ALGEBRAIC FORMULATION OF STATISTICAL MECHANICS 1
   - 1.1. Background orientation 1
   - 1.2. The algebras 3
   - 1.3. States of the system 4
   - 1.4. Symmetry groups 6
   - 1.5. The GNS construction 7
   - 1.6. Ergodicity and global decomposition 8
   - 1.7. Space translations 10
   - 1.8. Local dynamics 13
   - 1.9. The KMS condition 14
   - 1.10. The local Gibbs state 15
   - 1.11. The global Gibbs state and global dynamics 16

2. THE IDEAL FERMI GAS 20
   - 2.1. Introduction 20
   - 2.2. Configuration space 21
   - 2.3. Fock-Cook space 22
   - 2.4. The algebras 24
     - 2.4.1. The quantum fields 24
     - 2.4.2. The local algebras 24
     - 2.4.3. Local gauge transformations 25
     - 2.4.4. Local space translations 26
     - 2.4.5. The quasilocal algebras 27
   - 2.5. The Fock-Cook state 28
   - 2.6. The spin algebra 29
   - 2.7. Time translations 32
     - 2.7.1. Local time translations 32
     - 2.7.2. Global time translations 34
     - 2.7.3. Convergence of the local time translations 35
   - 2.8. Computation of the local Gibbs state 36
   - 2.9. The global Gibbs state 39
   - 2.10. The thermodynamic representations 43
   - 2.11. The Liouville equation 46
   - 2.12. The KMS condition 46

3. THE IDEAL BOSE GAS 48
   - 3.1. Introduction 48
   - 3.2. Configuration space and Fock-Cook space 50

3.3. The algebras 51
3.3.1. The boson fields 51
3.3.2. The algebras 53
3.4. Symmetries 54
3.5. The Fock-Cook state 55
3.6. Local dynamics 56
3.7. The local Gibbs state 57
3.8. The global Gibbs state 59
3.9. Global dynamics 67

4. THE BCS MODEL 70
4.1. Introduction 70
4.2. The model kinematics 71
4.3. The local Hamiltonian 72
4.4. The local Gibbs state 74
4.5. The global Gibbs state 76
4.6. The Haag-Bogoliubov Hamiltonian 80
4.7. The thermodynamic representation 82
4.8. Time translations 85

5. LATTICE GASES 89
5.1. Introduction 89
5.2. Spin-lattice kinematics 91
5.3. Spin-lattice dynamics 94
5.4. Spin-½ model dynamics 97
5.5. Ising model kinematics 99
5.6. Ising model dynamics and the transfer matrix 102
5.7. The global Gibbs state 104
5.8. Interlude: one dimension 105
5.9. Spontaneous breakdown of spin-reversal symmetry 105

BIBLIOGRAPHY 107

INDEX 117

1
The algebraic formulation of statistical mechanics

'What is lofty can be said in any language;
what is mean should be said in none.'
MIAMONIDES

1.1. Background orientation

SINCE the pioneering work of Boltzmann [1,2] and Gibbs [1] over seventy years ago, the generally agreed goal of statistical mechanics has been to derive the macroscopic properties of matter from a knowledge of the mechanics of the underlying microscopic entities (Ehrenfest and Ehrenfest [1]).

The guiding principle in such a derivation is to seek *less detailed information* at each stage in the transition from the microscopic to the macroscopic description. For it is only certain collective qualities of large systems which obey regular macroscopic laws; and then only if the total number of degrees of freedom N is large enough.

This principle would remain cardinal even if complete analytic solutions of the microscopic dynamical equations were available. As Gibbs pointed out (Gibbs [1], preface):

The laws of thermodynamics, as empirically determined, express the approximate and probable behaviour of systems of a great number of particles, or more precisely, they express the laws of mechanics for such systems as they appear to beings who have not the fineness of perception to enable them to appreciate quantities of the order of magnitude of those which relate to single particles, and who cannot repeat their experiments often enough to obtain any but most probable results.

The 'great number of particles' mentioned by Gibbs requires one to choose between constructing a theory with a large but finite number of degrees of freedom N contained within a finite volume V, or constructing one with both N and V actually infinite.

Infinite systems are an idealization, of course. But such systems allow precise descriptions of certain important properties, such as pure thermal equilibrium states, unaffected by any surface effects. By the principle of *quid pro quo* we must expect some resulting difficulty. These are technical mathematical problems associated with the infinite number of degrees of freedom. As they are analysed with the help of the theory of operator algebras, one speaks of *algebraic statistical mechanics.*

Infinite systems of interest in statistical mechanics are equipped with the dual notions of spatial localizability and strict local commutativity in the non-relativistic sense. This remark furnishes us with one way of constructing such systems. We start with well-defined quantities associated with finite N and V and take the infinite volume limit smoothly and in such a way as to preserve the intensive thermodynamic parameters. In particular, we must be careful to keep the temperature and density fixed. Historically, this limit is known as the thermodynamic limit. In this way one ends up with both local quantities, which can be associated with a finite volume, and quasilocal quantities, which pertain to the system as a whole and cannot be strictly localized.

Having come this far, one is tempted to seek a formulation of such quasilocal systems which is less dependent upon the finite volume subsystems than the formulation used in this book. Such matters are an open question, but there is reason to believe that thermal equilibrium states can be found directly by using the KMS condition, which we shall explain in § 1.9 below (Moya (1); Sewell (2)).

Several of the models discussed in this book have sharp phase transitions which are each associated with a sudden change of symmetry. This phenomenon is known as a spontaneous breakdown of symmetry. The natural and explicit treatment of systemic symmetries, especially to characterize phase transitions, is one of the main contributions of the algebraic formulation.

Another contribution by the algebraic formulation is the clarification of certain puzzling features of the dynamics of infinite systems. In the BCS model, for example, it was expected that a certain approximation to the true Hamiltonian would become exact in the limit, as it differed from the Hamiltonian by terms of the order of N^{-1} and smaller. Although obvious, this expectation is not true. This is one of the phenomena associated with the existence of *inequivalent representations* of operator algebras, and is peculiar to infinite systems.

That no fully satisfactory treatment of non-equilibrium phenomena exists so far is a great deficiency of all microscopic statistical mechanical theories. This deficiency is shared by the algebraic theory. One aspect of this problem is the question of the ergodicity of a system, which is known to be extraordinarily difficult to prove for realistic models. In the algebraic theory a conceptually precise framework for non-equilibrium phenomena exists, and a generalized master equation can be rigourously derived (Presutti, Scacciatelli, Sewell, and Wanderlingh (1)), although quantitative results on the decay to equilibrium are scarce (Davies (1,2); Emch and Radin (1); Lima and Verbeure (1,2); Radin (1,3,5,6)).

The following remarks about our notation may be helpful. For conceptual clarity, the distinction between a function and its images is maintained. We shall employ the mapping notation $f : A \to B$ to mean that the function f maps its domain A into its codomain B. The defining formula will be written either as $f(a) = b$, or $f : a \mapsto b$, whichever seems clearer. Certain standard symbols such as union, intersection, containment, etc. are used without our defining them.

In addition, certain less familiar concepts will be used without definitions. In particular we have in mind tensor products, direct products, direct sums and direct integrals, and inductive limits. These are defined and analyzed at various levels of abstraction in the following books: Auslander [1], Choquet [1], Dixmier [1,2], Gel'fand and Vilenkin [1], Grothendieck [1], Guichardet [1,2], Robertson and Robertson [1], Sakai [1], Segal and Kunze [1], and Sternberg [1].

1.2. The algebras

We shall now present an outline of the algebraic formulation of statistical mechanics.

Some general references are: Araki (1-3), Araki and Woods (1), Araki and Wyss (1), Dell'Antonio (1), Dubin and Sewell (1), Eckmann and Guenin [1], Emch [1], Haag and Kastler (1), Haag and Schroer (1), Haag, Hugenholtz, and Winnink (1), Robinson (2,3), Ruelle (1,2) [1], and Segal [1], (1,2,4). These are concerned mainly with the physical aspect. The mathematics used in this section may be found in Choquet [1], Dixmier [1,2], Guichardet [1,2], Robertson and Robertson [1], and Sakai [1].

One begins by declaring some finite-dimensional Euclidean space Γ to be the physical space of the model, i.e. the coordinate space within which the particles move or the lattice on which the spins vibrate.

Spatial localizability, a characteristic feature of statistical-mechanical systems, is associated with what are known as *local regions*. These are the bounded open subsets of Γ, sometimes required to satisfy certain additional regularity conditions, e.g. smooth boundaries. The set of local regions $\mathscr{L} = \{V\}$is assumed to contain a countable subfamily.† $\mathscr{M} = \{V_n : n \in \mathbb{N}\}$ which is ordered by inclusion and covers Γ. It follows that $\mathscr{M}$ is *absorbing* for $\mathscr{L}$: for any $V \in \mathscr{L}$ there is a least integer N such that $V \subset V_n$ for every $n \geqslant N$. Particular choices for $\mathscr{M}$ will be made at appropriate places in the succeeding chapters.

We next construct an underlying Hilbert space $\mathscr{H}$ appropriate for the model, on which the respective algebras can be concretely defined. For each $V \in \mathscr{L}$, moreover, we construct a Hilbert subspace $\mathscr{H}(V)$ of $\mathscr{H}$ such that $\mathscr{H}(V) \subset \mathscr{H}(W)$ when $V \subset W$. Upon abbreviating $\mathscr{H}(V_n)$ for $V_n \in \mathscr{M}$ by $\mathscr{H}_n$, these local Hilbert spaces are such that $\mathscr{H}$ is the Hilbert space inductive limit‡ of the $\{\mathscr{H}_n : n \in \mathbb{N}\}$; we write

$$\mathscr{H} = \varinjlim \{\mathscr{H}_n : n \in \mathbb{N}\} \tag{1.1}$$

for this process (Choquet [1]; Guichardet [1,2]).

There are local subalgebras $\mathscr{A}(V)$ associated with each $V \in \mathscr{L}$, appropriately chosen C^*- or W^*-subalgebras of $\mathbb{B}[\mathscr{H}(V)]$, the W^*-algebra of all bounded

† Our convention is to write $\mathbb{N} = \{1, 2, 3, \ldots\}$, $\mathbb{Z} = \{0, \pm 1, \pm 2, \ldots\}$, $\mathbb{R}$ = real numbers and $\mathbb{C}$ = complex numbers, $\mathbb{R}^+ = \{x \in \mathbb{R}: x \geqslant 0\}$.

‡ It is possible to write $\mathscr{H}$ as the closure of the union of the $\mathscr{H}_n$ as we shall do for the algebras in eqn (1.4).

operators on $\mathscr{H}(V)$.† The self-adjoint elements of $\mathscr{A}(V)$ are those observables of the system which can be localized within V. This clearly constrains the choice for $\mathscr{A}(V)$, and it can happen that there is not a best choice, e.g. for the Bose gas (cf. Kastler (1); Manuceau, Sirugue, Testard, and Verbeure (1); Manuceau (1); Wieringa (1), § 2.3; Slawney (1)).

In view of their construction, these subalgebras are mutually *isotonic*, meaning that $\mathscr{A}(V) \subset \mathscr{A}(W)$ when $V \subset W$, and $[\mathscr{A}(V), \mathscr{A}(U)]_- = 0$ when $V \cap U = \emptyset$‡. With the abbreviation $\mathscr{A}_n \equiv \mathscr{A}(V_n)$ for all $V_n \in \mathscr{M}$, the family $\{\mathscr{A}_n : n \in \mathbf{N}\}$ is inductive and so has a C^*-algebra as its C^*-inductive limit (Dixmier [1,2]; Guichardet [1,2]; Sakai [1]; Takeda (1)). This limiting algebra $\mathscr{A}$ is called the *quasilocal algebra* for the system:

$$\mathscr{A} = \lim_{\longrightarrow} \{\mathscr{A}_n : n \in \mathbf{N}\}. \tag{1.2}$$

It is also appropriate to define the systemic *local algebra* $\mathscr{A}_L$ by

$$\mathscr{A}_L = \cup_{V \in \mathscr{L}} \mathscr{A}(V) \tag{1.3}$$

This algebra is a norm-dense subalgebra of $\mathscr{A}$, i.e. there is a theorem (Sakai [1], proposition 1.23.2) to the effect that §

$$\mathscr{A} = \text{un. cl.}\,(\mathscr{A}_L). \tag{1.4}$$

The physical significance of these two algebras is that the self-adjoint elements of $\mathscr{A}$ (resp. $\mathscr{A}_L$) are the systemic observables (resp. localizable systemic observables). See Emch [1], Haag and Schroer (1), Haag and Kastler (1), and Segal [1] for a discussion of the physical significance of the axioms.

1.3. States of the system

In the algebraic formulation the concept of the *states of the system* has the following precise meaning: positive normalized linear functionals on the quasilocal algebra; states of the local subsystems are similarly defined on the local subalgebras. As regards notation, the set of states of any C^*- or W^*-algebra $\mathfrak{A}$ will be written $\mathfrak{S}(\mathfrak{A})$ and the value of any element $A \in \mathfrak{A}$ in the state $\psi \in \mathfrak{S}(\mathfrak{A})$ as $\psi(A)$ which is a complex number; the notation $\langle\psi; A\rangle$ is also found in the literature and emphasizes the duality between $\mathfrak{A}$ and $\mathfrak{S}(\mathfrak{A})$. (Dixmier [1,2]; Emch [1]; Ruelle (1,2), [1]; Sakai [1]).

† For any Hilbert space $\mathscr{H}$, $\mathbf{B}(\mathscr{H})$ is our symbol for the W^*-algebra of all bounded operators on it.

‡ $\emptyset$ is the empty set [,]_ is the commutator.

§ It is possible to define the C^*-inductive limit by (1.3) and (1.4); our definition is that in the Sakai text, namely constructing the free algebra on the topological product $\mathsf{X}_n \mathscr{A}_n$ and imposing isotony relations.

Most texts on quantum theory refer to vectors $v \in \mathcal{H}$ in the Hilbert space for the problem as states, which seemingly differs from our above definition. The connection between the two terminologies is that a vector $v \in \mathcal{H}$ determines a particular sort of state $\omega_v \in \mathfrak{S}(\mathbb{B}(\mathcal{H}))$ known as a *vector state* (Sakai [1], 2.7.7) through the formula

$$\omega_v(A) = (v, Av)_{\mathcal{H}} \qquad (\forall A \in \mathbb{B}(\mathcal{H})). \tag{1.5a}$$

But our definition is far more general; we include density matrices as states, for example. Recall that a *density matrix* δ is properly defined as a positive trace-class operator on $\mathcal{H}$, normalized in trace: $\mathrm{tr}(\delta) = 1$. The state $\phi \in \mathfrak{S}(\mathbb{B}(\mathcal{H}))$ in question is defined by

$$\phi(A) = \mathrm{tr}_{\mathcal{H}}\ (\delta A) \qquad (\forall A \in \mathbb{B}(\mathcal{H})), \tag{1.5b}$$

and is known as a *normal state* (Sakai [1], 1.15.3 - 1.15.6).

Note, in this regard, that for any density matrix δ there is an orthonormal basis $\{e_n : n \in \mathbb{N}\}$ of $\mathcal{H}$ with associated projection operators $\{P_n = e_n \otimes e^*_n : n \in \mathbb{N}\}$ and a family of eigenvalues, i.e. real numbers $\{\lambda_n \in [0,1]\,;\, n \in \mathbb{N}\}$, satisfying $\Sigma_{n\in\mathbb{N}}\ \lambda_n = 1$, for which

$$\delta = \sum_{n\in\mathbb{N}} \lambda_n P_n \tag{1.6}$$

in the strong sense. That is, for any $v \in \mathcal{H}$, $\| \delta(v) - \sum_{m=1}^{M} \lambda_m P_m(v) \| \to 0$ as $M \to \infty$ (Sakai [1], 1.15.4).

By convention, all Hilbert spaces shall be assumed to be separable and all algebras have an identity element $\mathbb{1} \in \mathfrak{A} : \mathbb{1} A = A \mathbb{1} = A$ for every $A \in \mathfrak{A}$. Any exceptions will be noted explicitly.

The above definitions of vector state and normal state will hold for any concrete C^*-algebra of operators on a Hilbert space, the only case we need. In any event, equivalent definitions, valid for any abstract C^*-algebra, are known.

There are, moreover, many states which are neither vector nor normal. One class of states of great importance to our work is the class of locally normal states. A state ψ is *locally normal* if its restrictions $\psi \restriction V$ to the local algebras $\mathcal{A}(V)$ are given through density matrices on the Fock-Cook spaces $\mathcal{H}(V)$ (Robinson (5)). In cases where the $\mathcal{A}(V)$ are W^*-algebras, this is equivalent to these restrictions of the state to each local subalgebra being *normal*, i.e. ultraweakly continuous (Sakai [1], 1.15.5). Such states are relevant for statistical mechanics because they roughly correspond to finite mean particle densities for every finite (i.e. local) region; however, the densities are not necessarily uniformly bounded with respect to volume. Zero densities correspond to normal states† (Chaiken (1,2);

† Not all locally normal states correspond to finite mean densities as the following counter-example shows. Assume the V-region number operator for bosons N_V is unbounded; this will be shown in Chapter 3. Consider the vector state ω_ψ where the vector ψ is *not* in the domain of N_V. Then the state ω_ψ is normal but $\omega_\psi\ (N_V/V)$ is not defined.

Dell'Antonio and Doplicher (1); Dell'Antonio, Doplicher, and Ruelle (1); Hugenholtz and Wieringa (1); Moya (1)).

Elementary computation shows that the normalization condition for states $\psi(\mathbb{1}) = 1$ implies that $\mathfrak{S}(\mathfrak{A})$ is a convex set. Furthermore, it is compact in the *weak*-topology*. Because $\mathfrak{S}(\mathfrak{A})$ is convex, an important role is played by the *extremal states*, i.e. the non-trivially indecomposable states. For extremal elements of convex sets are the convex analogue of basis vectors in a vector space. For historical reasons, extremal elements of $\mathfrak{S}(\mathfrak{A})$ are also known as pure states. The set of extremal states on $\mathfrak{A}$ shall be denoted $\mathfrak{E}(\mathfrak{A})$. (Choquet [1], §12; Dixmier [1,2]; Phelps [1]; Sakai [1], Chap. 3).

1.4. Symmetry groups

In addition to states and observables, the algebraic scheme encompasses symmetries in an intrinsic fashion. To say that a locally compact group $G = \{g\}$ is a symmetry group of the system means that G will act as a *group of automorphisms* both of the C^*-algebra $\mathfrak{A}$ and of the states $\mathfrak{S}(\mathfrak{A})$. (Dixmier [1,2]; Emch [1]; Ruelle [1]; Sakai [1]; Varadarajan [1]). For each group element $g \in G$ there is a mapping α_g taking every algebra element A into another algebra element $\alpha_g(A)$. Both the group and the algebraic structures are preserved (continuously in case of interest to us):

$$\alpha_g \,.\, \alpha_h(A) = \alpha_{(gh)}(A);\ \alpha_e(A) = A, \tag{1.7a}$$

where $e \in G$ is the group identity element. One also has

$$\begin{gathered}\alpha_g(A+B) = \alpha_g(A) + \alpha_g(B);\ \alpha_g(AB) = \alpha_g(A)\,.\,\alpha_g(B)\\ \alpha_g(A^*) = [\alpha_g(A)]^*.\end{gathered} \tag{1.7b}$$

Our notation for such an automorphism group shall be

$$\alpha : G \to \mathrm{Aut}\,(\mathfrak{A}), \tag{1.7c}$$

with images written either $\alpha_g(A)$ or $\alpha_g : A \mapsto \alpha_g(A)$ when an explicit expression for $\alpha_g(A)$ is available.

As far as the states are concerned, the symmetries act on them dually (synonyms: transposed, adjoint, contragrediently). The precise symmetry action is given by the definition

$$\begin{gathered}\alpha^* : G \to \mathrm{Aut}\,[\mathfrak{S}(\mathfrak{A})],\\ [\alpha_g^*\,\psi]\,(A) = \psi\,[\alpha_{g^{-1}}(A)]\end{gathered} \tag{1.8}$$

for every $g \in G$, $A \in \mathfrak{A}$, and $\psi \in \mathfrak{S}(\mathfrak{A})$. For notational brevity we shall sometimes write $\alpha(G)$ (resp. $\alpha^*(G)$) or even α (resp. α^*). When considering the quasilocal algebra $\mathscr{A}$ and its subalgebras, typical notations for automorphisms will be

$\alpha \in \text{Aut}\,[\mathscr{A}]$, $\alpha_V \in \text{Aut}\,[\mathscr{A}(V)]$, and $\alpha_n \in \text{Aut}\,(\mathscr{A}_n)$; the problem of whether or not—or in what sense—a formula such as $\lim_{n\to\infty} \alpha_n = \alpha$ holds will often be considered in succeeding chapters.

A state $\psi \in \mathfrak{S}(\mathfrak{A})$ will be said to be $\alpha(G)$-invariant or sometimes simply G-invariant if

$$\alpha_g{}^*(\psi) = \psi \tag{1.9}$$

for every $g \in G$. The set of G-invariant states will be written $\mathfrak{S}(\mathfrak{A}\,;G)$ or some abbreviation of this.

Just as $\mathfrak{S}(\mathfrak{A})$ is weak*-compact convex, $\mathfrak{S}(\mathfrak{A}\,;G)$ being a closed subspace of a compact space is compact in that topology, i.e. is weak*-compact (Choquet [1], Vol. I, p.23). Following Segal (2), the extremal G-invariant states are known as *G-ergodic states*. The set of G-ergodic states is written $\mathfrak{E}(\mathfrak{A}\,;G)$.

1.5. The GNS construction

In order to analyse $\mathfrak{S}(\mathfrak{A})$, we must now see how any state $\psi \in \mathfrak{S}(\mathfrak{A})$ can be canonically associated with a *representation* $\pi_\psi : \mathfrak{A} \to B(\mathscr{H}_\psi)$ on a certain Hilbert space $\mathscr{H}_\psi$, *cyclic* with respect to a distinguished vector $\Omega_\psi \in \mathscr{H}_\psi$. Having done so, we shall write $\psi \sim [\mathscr{H}_\psi, \pi_\psi, \Omega_\psi]$ and refer to $[\mathscr{H}_\psi, \pi_\psi, \Omega_\psi]$ as the GNS triple associated with ψ (Dixmier [1]; Emch [1]; Gel'fand and Naimark (1), Ruelle [1]; Sakai [1], 1.16, 1.21; Segal (1)).

The crux of the matter is that for any $\psi \in \mathfrak{S}(\mathfrak{A})$, it can be checked that $\psi(B^*A)$ is a *Hermitian form* on $\mathfrak{A}$; i.e. a sesquilinear positive semi-definite mapping $\mathfrak{A} \times \mathfrak{A} \to \mathbf{C}$. Consequently the pair $(\mathfrak{A}, \psi)$ is a pre-Hilbert space, which can be made into a Hilbert space in a standard way. Let $\mathfrak{B}_\psi = \{A \in \mathfrak{A} : \psi(A^*A) = 0\}$ denote the indicated closed left ideal (Dixmier [1,2]; Sakai [1]) of $\mathfrak{A}$, and $\mathfrak{A}/\mathfrak{B}_\psi$ the corresponding quotient space (Choquet [1]). Then the canonical Hilbert space $\mathscr{H}_\psi$ is defined to be the closure of this quotient space in the induced norm, and we write: $\mathscr{H}_\psi = \text{n.cl.}(\mathfrak{A}/\mathfrak{B}_\psi)$.

For every $A \in \mathfrak{A}$, let $[A]$ be its equivalence class in $\mathscr{H}_\psi$. Then the inner product in $\mathscr{H}_\psi$ is related to the original state through the formula

$$\psi(C^*D) = \langle [A], [B] \rangle, \tag{1.10}$$

where $C \in [A]$, $D \in [B]$ are any members of the respective equivalence classes.

The formula

$$\pi_\psi(A)\,[B] = [AB] \tag{1.11}$$

defines the required representation of $\mathfrak{A}$ on $\mathscr{H}_\psi$. To see that it is cyclic, define the distinguished vector as the equivalence class of the $\mathfrak{A}$-identity

$$\Omega_\psi \equiv [\mathbb{1}] \tag{1.12}$$

and note that $\mathfrak{A}/\mathfrak{B}_\psi$ is dense in $\mathscr{H}_\psi$ by definition. But it then follows that the following set equality holds:

$$\{\pi_\psi(A)\Omega_\psi : A \in \mathfrak{A}\} = \{[A] : A \in \mathfrak{A}\} = \mathfrak{A}/\mathfrak{B}_\psi .$$

This is just the definition of cyclicity, so π_ψ is a cyclic representation and Ω_ψ a cyclic vector.

Out of this we may extract the important formula

$$\psi(A) = \langle\Omega_\psi, \pi_\psi(A)\Omega_\psi\rangle, \tag{1.13}$$

which we shall often use.

The spaces $\mathscr{H}_\psi$ generally have little or no relation to the space $\mathscr{H}$ on which $\mathscr{A}$ is defined, for states $\psi \in \mathfrak{S}(\mathscr{A})$. In fact, Herman and Takesaki (1) and Takesaki (1) have shown that up to a technical condition of factor type, different temperatures and fugacities correspond to *unitarily inequivalent Hilbert spaces* associated with the respective thermal equilibrium states on $\mathscr{A}$. So, far from there being but one Hilbert space in the theory, it is the other extreme which presents itself.

Finally, let us note here that Hugenholtz and Wieringa (1) have shown that if $\psi \in \mathfrak{S}(\mathscr{A})$ is locally normal, $\mathscr{H}_\psi$ is *a priori* separable.

1.6. Ergodicity and global decomposition

With the following brief remarks we shall sketch the relation between symmetry groups and ergodic theory. The following references furnish more extensive discussions of this topic: Dixmier [1,2], Doplicher, Kadison, Kastler, and Robinson (1), Doplicher and Kastler (1), Doplicher, Kastler, and Størmer (1), Emch [1], Kastler, Haag, and Michel (1), Kastler, Mebkhout, Loupias, and Michel (1), Kastler and Robinson (1), Lanford and Ruelle (1), Radin (2,4,5); Robinson and Ruelle (1), Ruelle [1], (3), Sakai [1], Chap.3, Segal (1), Sewell (1), and Van Dongen and Verboven (1).

As we have already noted, extremal G-invariant states are also known as G-ergodic states. The reason for this name is that a certain non-commutative generalization of the classical ergodic theorem holds† for $\pi_\psi(\mathscr{A})$ on $\mathscr{H}_\psi$, every $\psi \in \mathfrak{E}(\mathscr{A}; G)$ (Arnold and Avez [1]).

First though, we note that if $a : G \to \mathrm{Aut}(\mathscr{A})$ is an automorphism group, and $\psi \in \mathfrak{S}(\mathscr{A}; G)$ a G-invariant state, the formula

$$[a_g(A)] \equiv U_\psi(g^{-1})[A] \qquad (\forall g \in G, A \in \mathscr{A}) \tag{1.14}$$

serves to define an isometry group $\{U_\psi(g) \in \mathbb{B}(\mathscr{H}_\psi) : g \in G\}$ on $\pi_\psi(\mathscr{A})\Omega_\psi = \{B\,\Omega_\psi : \forall B \in \pi_\psi(\mathscr{A})\}$.

It follows from this that U_ψ is extendable to a unitary group on $\mathscr{H}_\psi$. And if a is strongly continuous, then U_ψ is strongly continuous:

† In a standard notation, $\pi_\psi(\mathscr{A}) = \{\pi_\psi(B) : B \in \mathscr{A}\}$ is the image of $\mathscr{A}$ under the mapping π_ψ.

$$\lim_{g \to e} \| U_\psi(g) v - v \| = 0 \tag{1.15}$$

for every vector $v \in \mathscr{H}_\psi$. Starting from eqn (1.14), the corresponding action of G on the operators in $\pi_\psi(\mathscr{A})$ is given by the U_ψ - similarity transformation

$$\pi_\psi(a_g A) = U_\psi(g)\, \pi_\psi(A)\, U_\psi(g^{-1}). \tag{1.16}$$

Recall that $\Omega_\psi \in \mathscr{H}_\psi$ is the vector corresponding to the equivalence class of $\mathbb{1} \in \mathscr{A}$. But then

$$\begin{aligned} U_\psi(g)\,\Omega_\psi &= [a_{g^{-1}}(\mathbb{1})] \\ &= [\mathbb{1}] \\ &= \Omega_\psi \qquad (\forall g \in G); \end{aligned}$$

id est, the canonical cyclic vector is $U_\psi(G)$-invariant (or simply G-invariant):

$$U_\psi(g)\,\Omega_\psi = \Omega_\psi \qquad (\forall g \in G). \tag{1.17}$$

But there may be other G-invariant vectors in $\mathscr{H}_\psi$. In order to allow for this possibility, let $\mathbb{E}_\psi$, be the closed subspace of $\mathscr{H}_\psi$ which consists of all G-invariant vectors:

$$\mathbb{E}_\psi = \{v \in \mathscr{H} : U_\psi(g) v = v \,(\forall g \in G)\}. \tag{1.18}$$

An analogue of Von Neumann's ergodic theorem (Greenleaf [1]; Riesz and Sz-Nagy [1]) may be proven: if the subspace $\mathbb{E}_\psi$ is spanned by Ω_ψ alone, then $\psi \in \mathfrak{E}(\mathscr{A}; G)$ must be G-ergodic. Said otherwise, if $P_\psi \in \mathbb{B}(\mathscr{H}_\psi)$ is the $\mathbb{E}_\psi$-projection operator: $P_\psi(\mathscr{H}_\psi) = \mathbb{E}_\psi$, then in this case P_ψ is of one-dimensional range, and hence $P_\psi = \Omega_\psi \otimes \Omega_\psi{}^*$. The proof may be found, e.g. in Sakai [1], proposition 3.1.10, cf. Emch [1]; Ruelle [1].

Lanford and Ruelle (1) have found the condition for the converse; they have termed it the *G-abelian property*. The algebra $\mathscr{A}$ is said to be $a(G)$-abelian if

$$\{P_\psi\, \pi_\psi(A)\, P_\psi \;:\; A \in \mathscr{A}\}$$

is an abelian family of operators for every ergodic state $\psi \in \mathfrak{E}(\mathscr{A}; G)$.

If $\mathscr{A}$ is $a(G)$-abelian, they showed that $\psi \in \mathfrak{E}(\mathscr{A}; G)$ iff $\mathbb{E}_\psi$ is one-dimensional. See either Emch [1] or Ruelle [1] for a proof.

Other forms of G-abelianess can be defined. Suppose, for instance, that there exists a sequence of group elements $\{g_n \in G : n \in \mathbb{N}\}$, not necessarily convergent, such that the limit

$$\lim_{n \to \infty} \| [a_{g_n}(A), B]_- \| = 0 \tag{1.19}$$

exists for every $A, B \in \mathscr{A}$; then $\mathscr{A}$ is said to be *asymptotically abelian* (*with respect to* $a(G)$). The connection with the ergodic theorem is that if $\mathscr{A}$ is asymptotically abelian, then it is G-abelian (proof: Sakai [1], 3.1.16).

We do not have the space to discuss any global decomposition theory. Suffice it to say that a typical theorem in this theory, of the sort of interest to us, is to prove that some G-invariant state $\psi \in \mathfrak{S}(\mathscr{A}; G)$ can be uniquely (or not) decomposed into G-ergodic states:

$$\psi = \int_{\mathfrak{E}(\mathscr{A};\,G)}^{\oplus} \varphi \, \mathrm{d}m(\varphi), \tag{1.20}$$

where m is a Radon probability measure Baire-concentrated on the ergodic states (Choquet [1]; Phelps [1]). This expression implies that in the cases of interest to us, the GNS triple $\psi \sim [\mathscr{H}_\psi, \pi_\psi, \Omega_\psi]$ decomposes analogously, e.g. $\mathscr{H}_\psi = \int^{\oplus} \mathscr{H}_\varphi \, \mathrm{d}m(\varphi)$. The construction of direct integrals is discussed in many texts, e.g. Gel'fand and Vilenken [1]. But note the following technical point. The decomposition of $\mathscr{H}_\psi$ does not generally follow from (1.20); the trivial group $G = \{e\}$ furnishes a counterexample. For if $\psi(A) = \mathrm{tr}(A)$ on the space $\mathbb{C}^2$, $\mathscr{H}_\psi$ is four-dimensional, $\mathscr{H}_\psi \simeq \mathbb{C}^4$. But the right-hand side of (1.20) is $\int_{\|v\| = 1} \langle Av, v\rangle$ and is a direct integral over an infinite number of spaces, i.e. $\int^{\oplus}_{\|v\| = 1} \mathscr{H}_v$ is infinite-dimensional. When the decomposition (1.20) is also the central decomposition, the GNS decomposition will follow.† Other cases, such as KMS decompositions (see § 1.9 below), which we use are usually favourable cases (Van Dongen and Verboven (1)).

Now suppose that eqn (1.20) holds, and that in addition there is another automorphism $\gamma \in \mathrm{Aut}(\mathscr{A})$ of $\mathscr{A}$ (or automorphism group $\gamma : G' \to \mathrm{Aut}(\mathscr{A})$). It can happen that $\psi \in \mathfrak{S}(\mathscr{A}; \gamma)$ is γ-invariant, but the G-ergodic components are not: $\varphi \notin \mathfrak{S}(\mathscr{A}; \gamma)$.

Even more to the point, suppose that ψ in (1.20) is a thermal equilibrium state. It sometimes is the case that for temperatures above some critical one, $T > T_c$, both ψ_T and the $\{\varphi_T\}$ are γ-invariant; but that for $T \leqslant T_c$, ψ_T is, yet the $\{\varphi_T\}$ are not γ-invariant. We say that ψ_T undergoes a *spontaneous breakdown of γ-symmetry associated with its G-ergodic decomposition.*

The physical interpretation of this is the following. There is reason to believe that ergodic states represent pure thermodynamic phases of the system provided they are thermal equilibrium states. But then the symmetry breakdown described above signals a phase transition; even more, it gives information concerning the contrasting symmetry properties of the phases.

1.7. Space translations

An important example of a systemic symmetry is the additive group Γ of physical space, with vector addition as the abelian group operation: $(\xi, \eta) \mapsto \xi + \eta$, with $\xi, \eta \in \Gamma$. As we shall show, the localizability property of the system enables Γ to give rise to a group of automorphisms $\sigma : \Gamma \to \mathrm{Aut}(\mathscr{A})$ of the quasilocal algebra.

First, however, we must go back and be somewhat more pedantic about a point already discussed, namely isotony. When we actually come to construct the underlying Hilbert spaces $\{\mathscr{H}(V) : V \in \mathscr{L}\}$ (resp. algebras $\{\mathscr{A}(V) : V \in \mathscr{L}\}$)

† This example is due to W. Wils and communicated privately by E. B. Davies.

for the models, they will be constructed as separate spaces (resp. algebras) in their own right, one for each $V \in \mathscr{L}$, rather than as subspaces of $\mathscr{H}$ (resp. $\mathscr{A}$). This being so, when we say, for example, that $\mathscr{H}(V)$ is a subspace of $\mathscr{H}(W)$ for $V \subset W$, we mean that there is a mapping of $\mathscr{H}(V)$ into $\mathscr{H}(W)$, which identifies $\mathscr{H}(V)$ with a subspace of $\mathscr{H}(W)$, etc. We shall use the following notation in this regard. For the local subsystems, we write

$$i(V,W) : \mathscr{H}(V) \mapsto \mathscr{H}(W) \tag{1.21a}$$

$$(\text{for } V \subset W;\, V, W \in \mathscr{L}).$$

$$j(V,W) : \mathscr{A}(V) \mapsto \mathscr{A}(W) \tag{1.22a}$$

For the injection of the subsystems into the quasilocal system, we write

$$i(V) : \mathscr{H}(V) \mapsto \mathscr{H} \tag{1.21b}$$

$$j(V) : \mathscr{A}(V) \mapsto \mathscr{A} \quad (V \in \mathscr{L}). \tag{1.21a}$$

It proves convenient to abbreviate $i(V_n, V_m)$ (resp. $j(V_n, V_m)$) by i_{nm} (resp. j_{nm}) for $V_n, V_m \in \mathscr{M}$, and $i(V_n)$ (resp. $j(V_n)$) by i_n (resp. j_n). The various i and j mappings are known as *injective mappings*, or simply *injections* (Choquet [1], Vol. I, p.8; Robertson and Robertson [1], p.88).

Returning to the problem of *space translations*, we assume—and this assumption will be true for all our models—that every local Hilbert space $\mathscr{H}(V)$ is naturally identifiable with every local Hilbert space $\mathscr{H}(V + \xi)$, where $\xi \in \Gamma$ and the set $V + \xi$ is the local region $\{\xi + \eta : \eta \in V\}$.

By natural we mean unitarily equivalent, i.e. there exists a unitary operator

$$S_V(\xi) : \mathscr{H}(V) \to \mathscr{H}(V + \xi) \quad (\forall V \in \mathscr{L}, \xi \in \Gamma) \tag{1.23a}$$

effecting the identification. One cannot say that $\{S_V(\xi) : \xi \in \Gamma\}$ is a unitary group on $\mathscr{H}(V)$ even though an additive law of composition holds: $S_{V+\xi}(\eta)\, S_V(\xi) = S_V(\xi + \eta)$, because $S_{V+\xi}$ acts on $\mathscr{H}(V + \xi)$, whereas S_V acts on $\mathscr{H}$, as we shall see.

If this scheme is to be consistent, the translations must be compatible with the injections, which is where localizability and isotony enter. They are compatible, and the compatibility condition for Hilbert spaces is the mapping identity

$$i(V+\xi, W+\xi) \circ S_V(\xi) = S_W(\xi) \circ i(V, W), \tag{1.24}$$

which is easily understood from the corresponding *commutative diagram* (Spanier [1]) seen in Fig. 1.

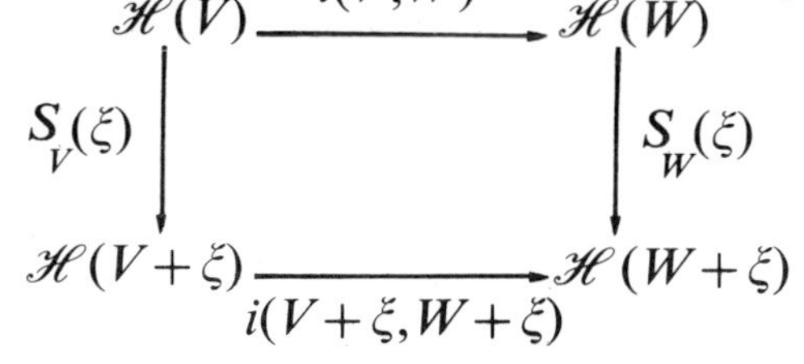

FIG. 1

Knowing that the scheme is consistent, we may use the $\{S_V : V \in \mathscr{L}\}$ to define a translation unitary group on $\mathscr{H}$. As $\mathscr{H} = \varinjlim (\mathscr{H}_n)$ is an inductive limit, the required unitary group is the inductive limit of the corresponding mappings (Choquet [1]):

$$S(\xi) = \varinjlim \{S_{V_n}(\xi)\} \quad (\forall \xi \in \Gamma). \tag{1.23b}$$

We shall write $S(\Gamma)$ for the set of images; the group property is represented by the composition law $S(\xi) S(\eta) = S(\xi + \eta)$. Although we shall not make use of the fact, S is strongly continuous on $\mathscr{H}$: $\| S(\xi) v - v \| \to 0$, as $\xi \to 0$ in Γ, for every vector $v \in \mathscr{H}$. Consequently Stone's theorem applies: there is a self-adjoint momentum operator Π on $\mathscr{H}$, defined through $S(\xi) = \exp(\mathrm{i}\, \xi\, \Pi)$ (Kato [1]).

This transformation scheme for the Hilbert spaces induces a corresponding one for the algebras. For the local subalgebras we define

$$\sigma_V(\xi) : \mathscr{A}(V) \to \mathscr{A}(V + \xi) \tag{1.24a}$$

by the formula

$$\sigma_V(\xi) : A \to S_V(\xi) A\, S_V(-\xi) \quad (A \in \mathscr{A}(V), \xi \in \Gamma). \tag{1.24b}$$

And upon defining the quasilocal algebra automorphism group

$$\sigma : \Gamma \to \operatorname{Aut}(\mathscr{A}) \tag{1.24c}$$

as the inductive limit mapping

$$\sigma = \varinjlim \{\sigma_{V_n}\}, \tag{1.24d}$$

the explicit formula

$$\sigma(\xi)[B] = S(\xi) B S(-\xi) \quad (B \in \mathscr{A}, \xi \in \Gamma) \tag{1.24e}$$

for σ as a similarity transform follows.

It is a characteristic property of the models we shall consider that they are $\sigma(\Gamma)$-asymptotically abelian:

$$\lim_{|\xi| \to \infty} \| [\sigma(\xi) A, B]_- \| = 0 \quad (\forall A, B \in \mathscr{A}), \tag{1.25}$$

where $|\xi|$ is the Euclidean norm of $\xi \in \Gamma$. For a proof see Emch [1] and Ruelle [1]. Physically, this result says that one can always reduce the effect of any observable within one's laboratory to any required accuracy by translating the laboratory far enough. Mathematically it implies, amongst other things, that for any Γ-invariant state $\psi \in \mathfrak{S}(\mathscr{A}; \Gamma)$ there is a unique (maximal) measure M_ψ depending upon ψ, concentrated on the Γ-ergodic states, $M_\psi [\mathfrak{S}(\mathscr{A}; \Gamma) \backslash \mathfrak{E}(\mathscr{A}; \Gamma)] = 0$, up to exceptional Baire subsets, and we write

$$\psi = \int^{\oplus}_{\mathfrak{E}(\mathscr{A}; \Gamma)} \varphi \, \mathrm{d}M_\psi(\varphi). \tag{1.26}$$

References to this decomposition are in the previous section. Sakai ([1], Chap. 3) gives the general theory, Emch [1] and Ruelle [1] relate it to the case at hand, and Sewell (2) relates it to the dynamics.

1.8. Local dynamics

In the previous section we showed how space translations could be accommodated in the algebraic theory as a group of automorphisms of the quasilocal algebra. We would like to do the same for the dynamics of the system, but because of certain technical difficulties this is not always possible (Dubin and Sewell (1); Hugenholtz and Weiringa (1); Ruskai (1); Sirugue and Winnink (2)).

There would seem to be three cases to consider.† The most favourable, which is pertinent to the ideal Fermi gas and certain lattice spin models, can be treated in direct analogy with the space translations. The dynamical evolution of the system is given by a one-parameter group of automorphisms $\tau : \mathbb{R} \to \mathrm{Aut}\,(\mathscr{A})$ of the quasilocal algebra; the parameter $t \in \mathbb{R}$ is physically interpreted as the time (cf. Haag, Hugenholtz, and Winnink (1)).

The least favourable case considered in this book is that of the ideal Bose gas model. If ϕ_β denotes the thermal equilibrium state on the pertinent quasilocal algebra $\mathscr{A}$ at inverse temperature $\beta = (kT)^{-1}$ and $\phi_\beta \sim [\mathscr{H}_\beta, \pi_\beta, \Omega_\beta]$; and if $\pi_\beta(\mathscr{A})'' \equiv \mathscr{A}_\beta''$ is the bicommutant ‡ and hence the weak closure (Sakai [1], 1.20) of the representing algebra $\pi_\beta(\mathscr{A}) \equiv \mathscr{A}_\beta$, the dynamics associated with thermal equilibrium are given by an automorphism group $\tau_\beta : \mathbb{R} \to Aut\,(\mathscr{A}_\beta'')$. States not associated with the $\{\mathscr{H}_\beta : \beta \in \mathbb{R}^+\}$ must be examined in individual cases by considering the convergence properties of the local time automorphisms $\tau_V : \mathbb{R} \to \mathrm{Aut}\,[\mathscr{A}(V)]$ which exist in all cases: we consider only mechanistic systems.

The median situation, which shall be described here, will, it is hoped, prove to be the proper formulation of the dynamical problem for equilibrium cases. This situation is represented in this book by the BCS model. The conclusions are much as for the least favourable cases, i.e. an automorphism of $\mathscr{A}_\beta''$ results for thermal equilibrium situations. But we require the technical assumption that the local subalgebras be W^*-algebras, i.e. weakly closed: $\mathscr{A}(V)'' = \mathscr{A}(V)$ for every $V \in \mathscr{L}$. In view of the fact that the situation is sometimes more and sometimes less favourable in particular models, our remarks here are to be understood *mutatis mutandis.*

The systems under consideration are mechanistic, as we said, which means that for each $V_n \in \mathscr{M}$ the systemic dynamics is governed by a strongly continuous one-parameter unitary group $U^{(n)}$ generated by a Hamiltonian $\bar{H}_n$ (Helmberg [1];

† This is optimistic; physically interesting non-equilibrium states, such as those representing turbulence, could well be less favourable than the cases treated here.

‡ If $\mathfrak{A} \subset \mathbf{B}(\mathscr{H})$, then the commutant is defined to be the set $\mathfrak{A}' \equiv \{B \in \mathbf{B}(\mathscr{H}) : [A,B]_- = 0, \forall A \in \mathfrak{A}\}$ and $\mathfrak{A}'' \equiv (\mathfrak{A}')'$.

Hille and Phillips [1]; Kato [1]; Riesz and Sz-Nagy [1]). Thus

$$\begin{aligned} U^{(n)} &: \mathbb{R} \to \mathbb{B}(\mathscr{H}_n), \\ U^{(n)}(t) &= \exp(\mathrm{i}t\bar{H}_n). \end{aligned} \tag{1.27}$$

It is worth noting that $\bar{H}_n$ is the reduced Hamiltonian, implying that we shall be working within the grand canonical scheme. If μ_n denotes the chemical potential for V_n, and N_n the corresponding particle number operator, then in models for which it is relevant, there is a true Hamiltonian H_n which is related to the reduced Hamiltonian through the formula

$$\bar{H}_n = H_n - \mu_n N_n.$$

As the V_n are open sets, it is assumed that some self-adjoint extension has been chosen for the Hamiltonian: $(\bar{H}_n)^* = (\bar{H}_n)$, and that $\bar{H}_n$ is lower bounded (this on physical grounds) (Kato [1]).

The time-translation automorphism for the local region V_n, written $\tau_{V_n} \equiv \tau^{(n)}$, is implemented by the unitary group $U^{(n)}(\mathbf{R})$ defined in (1.27); the defining formula is

$$\begin{aligned} \tau^{(n)} &: \mathbb{R} \to \mathrm{Aut}(\mathscr{A}_n), \\ \tau_t^{(n)} &: A \mapsto U_t^{(n)} A\, U_{-t}^{(n)} \quad (t \in \mathbb{R}, A \in \mathscr{A}_n), \end{aligned} \tag{1.29}$$

i.e. $U^{(n)}$ acts by similarity transformation. We assume, furthermore, that $U^{(n)}(t) \in \mathscr{A}_n$ for every $t \in \mathbb{R}$; one says that $\tau^{(n)}$ is *inner* in such cases (Dixmier [1,3]; Sakai [1]). Before proceeding with the study of the limit of the $\tau^{(n)}$ $(n \to \infty)$, we examine the characteristic property of thermal equilibrium states, and define the local Gibbs states.

1.9. The KMS condition

It turns out that it is not time-translation invariance which characterizes thermal equilibrium. As Haag, Hugenholtz, and Winnink (1) (cf. Dubin and Sewell (1); Kastler, Pool, and Poulson (1); Takesaki [1]) first pointed out, it is states which satisfy the so-called KMS (Kubo (1); Martin and Schwinger (1)) condition which are to be interpreted as thermal equilibrium states.

Our definition of the KMS condition is more general than we shall actually need for time translations, but is no more complicated. Let $\psi \in \mathfrak{S}(\mathfrak{A})$, where $\mathfrak{A}$ is any C^*-algebra with identity, let $\beta \in \mathbb{R}^+$ be a real number, and $a : \mathbb{R} \to \mathrm{Aut}(\mathfrak{A})$ be a strongly continuous group of automorphisms.

For every two elements P and Q of $\mathfrak{A}$, form the two complex-valued functions $f, g : \mathbb{R} \to \mathbb{C}$ defined by

$$\begin{aligned} f(t) &= \psi\,[a_t(P)\,Q], \\ g(t) &= \psi\,[Q\, a_t(P)]. \end{aligned} \tag{1.30}$$

We say that ψ satisfies the KMS condition with respect to $[\beta, \alpha(\mathbb{R})]$ if there exist two complex functions $F, G : \mathbb{C} \to \mathbb{C}$ respectively

(1) analytic in the strips $\mathrm{Im}(z) \in (-\beta, 0)$ and $\mathrm{Im}(z) \in (0, +\beta)$;
(2) continuous on their boundaries;
(3) having limits, in the usual topology on $\mathscr{S}'(\mathbb{R})$:

$$\mathscr{S}'(\mathbb{R}) - \lim_{\epsilon \to 0^-} F(t + i\epsilon) = f(t)$$

$$\mathscr{S}'(\mathbb{R}) - \lim_{\epsilon \to 0^+} G(t + i\epsilon) = g(t); \tag{1.31}$$

(4) and whose Fourier transforms satisfy

$$\tilde{f}(\omega) = e^{\beta\omega} \tilde{g}(\omega) \tag{1.32}$$

in $\mathscr{S}'(\mathbb{R})$.

In a heuristic sense, (1.31) and (1.32) are sometimes written as

$$\psi\,[\alpha_t(P)\,Q] = \psi\,[Q\,\alpha_{t-i\beta}(P)].$$

Here $\mathscr{S}'(\mathbb{R})$ is the Schwartz space of tempered distributions over $\mathbb{R}$ (Gel'fand and Shilov [1,2]; Streater and Wightman [1]; Treves [1]).
For further discussion pertaining to the KMS condition, see: Araki (4), Araki and Miyata (1), Emch and Knops (1), Emch, Knops, and Verboven (1-3), Hugenholtz (1), Rocca, Sirugue, and Testard (1,2), and Sirugue and Winnink (1).

1.10. The local Gibbs state

Assume that the reduced Hamiltonian $\bar{H}_n$ is such that the semi-group (Helmberg [1]; Hille and Phillips [1]; Kato [1]) ($\beta = 1/kT$ is the inverse temperature)

$$\sigma_\beta^{(n)} = \exp(-\beta \bar{H}_n) \quad (\beta \in \mathbb{R}_*^+) \tag{1.33}$$

is of trace-class on $\mathscr{H}_n$; we shall have to prove this for each of the models, of course.† Then the grand partition function for $\mathscr{A}_n$ is well defined, and given by the formula

$$Z_\beta^{(n)} = \mathrm{tr}_n\, \sigma_\beta^{(n)}, \tag{1.34}$$

where tr_n denotes the unique trace on $\mathscr{H}_n$.

The Gibbs state or grand canonical state $\phi_\beta^{(n)} \in \mathfrak{S}(\mathscr{A}_n)$ for the region V_n is defined to be

$$\phi_\beta^{(n)}(A) = E_\beta^{(n)}(A) \,/\, E_\beta^{(n)}(\mathbb{1}) \quad (A \in \mathscr{A}_n), \tag{1.35}$$

† $\mathbb{R}_*^+ = \{x \in \mathbb{R} : x > 0\}$.

where

$$E_\beta^{(n)}(A) = \operatorname{tr}_n(\sigma_\beta^{(n)} A); \tag{1.36}$$

of course, $E_\beta^{(n)}(\mathbb{1}) = Z_\beta^{(n)}$ must be non-zero. Another way of writing this is to define the operator

$$\delta_\beta^{(n)} = \sigma_\beta^{(n)} / Z_\beta^{(n)}; \tag{1.37}$$

our assumption that $\sigma_\beta^{(n)}$ is trace-class implies that $\delta_\beta^{(n)}$ is a density matrix. Since $\phi_\beta^{(n)}(A) = \operatorname{tr}_n(\delta_\beta^{(n)} A)$, $\phi_\beta^{(n)}$ is by definition a normal state.

We shall not give a proof that $\phi_\beta^{(n)}$ is $[\beta, \tau^{(n)}]$ -KMS; one may find that in several places; the original proof is due to Haag, Hugenholtz, and Winnink (1). An expository account is in Winnink (1). It is easy to show that $\phi_\beta^{(n)}[\tau_t^{(n)}(A)B]$ is equal to $\phi_\beta^{(n)}[B\,\tau^{(n)}{}_{t-i\beta}(A)]$ formally, with certain assumptions, e.g. $\tau^{(n)}{}_{t-i\beta}$ is well defined for complex arguments, and $\bar{H}_n$ has a simple discrete spectrum with the eigenvectors forming a basis of $\mathscr{H}_n$. That is, $\{e_p : p \in \mathbf{Z}\}$ is an orthonormal basis for $\mathscr{H}_n$ with

$$\bar{H}_n e_p = \lambda_p e_p \quad (p \in \mathbf{Z}) \tag{1.38}$$

as the eigenvalue equations. As these assumptions will be true for the models in this book, let us continue with them. Then

$$\phi_\beta^{(n)}[\tau_t^{(n)}(A)B] =$$
$$= [Z_\beta^{(n)}]^{-1} \sum_{p,\,q \in Z} (e_p, Ae_q)(e_q, Be_p) \exp\,[it\lambda_p - \lambda_q) - \beta\lambda_p] \tag{1.39a}$$

and

$$\phi_\beta^{(n)}[B\tau_{t-i\beta}^{(n)}(A)] = \tag{1.39b}$$
$$= [Z_\beta^{(n)}]^{-1} \sum_{l,m \in Z} (e_l, Be_m)(e_m, Ae_l) \exp\,[it(\lambda_m - \lambda_l) - \beta\lambda_l + \beta(\lambda_l - \lambda_m)],$$

from which the equality is obvious. Of course, this is no proof, but it contains the essence of the matter; we shall assume that the technically minded reader will go to the references cited, and to Kastler, Pool, and Poulsen (1) as well.

1.11. The global Gibbs state and global dynamics

This section concerns the limit, as n approaches infinity, of $\phi_\beta^{(n)}$ and $\tau^{(n)}$; in the case of $\phi_\beta^{(n)}$, the constraint that the particle density $\phi_\beta^{(n)}[N_n / |V_n|] \equiv \rho_\beta^{(n)}$ associated with the region $V_n \in \mathscr{M}$ (of volume $|V_n|$) be prescribed must be levied. For all the models that we shall consider, the prescription shall be for uniform density: $\rho_\beta^{(n)} \equiv \bar{\rho} \in \mathbf{R}^+$, independent of temperature and volume. This is an essential ingredient of the grand canonical formalism we propose to employ.

It sometimes happens that the $\tau^{(n)}$ converge; for the ideal Fermi gas we shall show that there exists an automorphism group

$$\tau : \mathbf{R} \to \operatorname{Aut}(\mathscr{A}) \tag{1.40a}$$

of the quasilocal algebra which is the uniform limit of the local automorphisms, namely,

$$\| \tau_t^{(n)} A - \tau_t A \| \to 0 \quad (\forall A \in \mathscr{A}_L). \tag{1.40b}$$

The automorphism group $\tau(\mathbf{R})$ defined in this formula is continuous in t and extends continuously from the local algebra $\mathscr{A}_L$ to its uniform closure $\mathscr{A}$. In such an instance, one would separately prove the existence of a quasilocal Gibbs state $\phi_\beta \in \mathfrak{S}(\mathscr{A})$:

$$| \phi_\beta (A) - \phi_\beta^{(n)} (A) | \to 0 \quad (\forall A \in \mathscr{A}_L), \tag{1.41}$$

again extending from $\mathscr{A}_L$ to $\mathscr{A}$ by continuity. This is the most favourable case.

But sometimes eqn (1.40) is demonstrably untrue, as for the ideal Bose gas and the BCS model. Although there is some indication that the Bose gas can be treated as a median case, in this book we use the median case, which will be described now, as an example, and prove the necessary theorems for the Bose gas directly.

For the median case (BCS model), the convergence of the $\phi_\beta^{(n)}$ and the $\tau^{(n)}$ can be combined to prove an existence theorem associated with ϕ_β. In order to do so, we must assume that the following two convergence criteria are satisfied:

$$\lim_{n\to\infty} \phi_\beta^{(n)} \left[\prod_{j=1}^{k} \tau_{t_j}^{(n)} (A_j) \right] \text{ exists for all } A_1, \ldots, A_k \in \mathscr{A}_L, \; \forall\, t_1, \ldots, t_k \in \mathbf{R}; k < \infty; \tag{1.42}$$

$$\lim_{m\to\infty} \lim_{n\to\infty} \phi_\beta^{(n)} \left[\prod_{j=1}^{k} \tau_{t_j}^{(n)} (A_j) \prod_{p=k+1}^{d} \tau_{t_p}^{(m)} (A_p) \right] \text{ exists and is}$$

equal to

$$\lim_{n\to\infty} \phi_\beta^{(n)} \left[\prod_{j=1}^{d} \tau_{t_j}^{(n)} (A_j) \right] \text{ for all } A_1, \ldots, A_d \in \mathscr{A}_L; \; \forall\, t_1, \ldots, t_d \in \mathbf{R}; k, d < \infty. \tag{1.43}$$

We shall refer to these two crucial assumptions as DS I and II respectively (Dubin and Sewell (1)).

On the basis of these assumptions one may prove that a limiting global Gibbs state $\phi_\beta \in \mathfrak{S}(\mathscr{A})$ exists

$$\phi_\beta (A) = \lim_{n\to\infty} \phi_\beta^{(n)} (A) \quad (\forall A \in \mathscr{A}_L). \tag{1.44}$$

If ϕ_β is associated with $(\mathscr{H}_\beta, \pi_\beta, \Omega_\beta)$ by the GNS construction, it will be very convenient to write, as in §1.8,

$$\mathscr{A}''_\beta = [\pi_\beta(\mathscr{A})]'' \tag{1.45}$$

for the weak closure of the representing algebra; and $\Phi_\beta \in \mathfrak{S}(\mathscr{A}''_\beta)$ for the unique continuous extension of ϕ_β to $\mathscr{A}''_\beta$, defined by

$$\Phi_\beta (Q) = (\Omega_\beta, Q\, \Omega_\beta) \quad (\forall Q \in \mathscr{A}''_\beta). \tag{1.46}$$

Hence Φ_β is normal on the W^*-algebra $\mathscr{A}''_\beta$, an observation which enables one to apply a number of theorems arising from the KMS condition.

From the above convergence conditions it then follows that

$$\lim_{n\to\infty} \pi_\beta(\tau_t^{(n)} A) = U_\beta(t)\, \pi_\beta(A)\, U_\beta(-t), \tag{1.47}$$

where the strongly continuous one-parameter global unitary group U_β defines the one-parameter global time-translation automorphism group

$$\tau_\beta : \mathbf{R} \to \mathrm{Aut}\,(\mathscr{A}''_\beta) \tag{1.48a}$$

on $\mathscr{A}''_\beta$ in the familiar manner:

$$\tau_\beta(t)\,[Q] = U_\beta(t)\, Q\, U_\beta(-t) \tag{1.48b}$$

for every $Q \in \mathscr{A}''_\beta$ and $t \in \mathbf{R}$. Note that this is not an automorphism group of the quasilocal algebra $\mathscr{A}$ but of the weak closure of its Gibbs-state representative $\mathscr{A}''_\beta$. Moreover,

$$U_\beta(t)\, \Omega_\beta = \Omega_\beta, \tag{1.49}$$

so that Φ_β is (β, τ_β)-invariant. Even more, Φ_β is (β, τ_β)-KMS, and all that that implies.

It sometimes happens that Φ_β is the only (β, τ_β)-KMS ergodic state, as in the ideal Fermi gas, for instance. In fact, this is so for all inverse temperatures in that model. As previously mentioned, we interpret this as associating a unique pure thermodynamic equilibrium phase with Φ_β.

We might note that the Gibbs states for different temperatures lead to *disjoint* GNS representations, provided only that that at least one of the representations leads to a *Type-III factor* (Dixmier [1,2]; Emch [1]; Sakai [1]; Takesaki (1)), and hence to unitarily inequivalent Hilbert spaces (cf. §1.5).

Another satisfactory consequence of all this is that any KMS state must necessarily be locally normal, i.e. have the local finite mean density property (Takesaki and Winnink (1)).

A model, in our terms, consists of a quasilocal algebra acting on an underlying Hilbert space built over a configuration space. An explicit expression for the local Hamiltonians must be given. The model is solvable, in the minimal sense we employ in this book, if the explicit time evolution τ_β and the Gibbs state ϕ_β are known. Further, one must be able to give the KMS-ergodic decomposition for Φ_β and the GNS construction in explicit form, both for Φ_β and its ergodic components. We shall not examine the fine points of the overlap between the extremal KMS decomposition and extremal decompositions with respect to the amenable subgroups of the space translation group. Nor can we characterize any large class of non-equilibrium states for which a satisfactory dynamical framework can be given.

This ends our discussion of the general structure of the theory; our next task is to apply it to the models, starting with the ideal Fermi gas.

In summary, we have a physical space Γ and its local regions $\mathscr{L}$; an underlying Hilbert space $\mathscr{H}$ with local subspaces $\{\mathscr{H}(V) : V \in \mathscr{L}\}$; local C^*-subalgebras $\mathscr{A}(V) \subset \mathbb{B}[\mathscr{H}(V)]$ and quasilocal algebra $\mathscr{A} = \varinjlim (\mathscr{A}_n)$.

For each $V_n \in \mathscr{M}$ there is a reduced Hamiltonian $\bar{H}_n$ which generates the local time translations by similarity transformation with the unitary group $U_t^{(n)} = \exp(it\bar{H}_n)$; and the local Gibbs states through the semi-group $\sigma_\beta^{(n)} = \exp(-\beta\bar{H}_n)$.

Our main task is to evaluate the local Gibbs states and consider the thermodynamic limit of $\tau^{(n)}$ and $\phi_\beta^{(n)}$. Once these limits are found, we analyse them for spontaneous symmetry breakdowns, which would signal the onset of a phase transition.

As a final remark in this connection, let us point out that these systems are infinite and so it would be more satisfactory to define certain properties, e.g. the thermal equilibrium state, globally, rather than as the limit of local quantities. For the equilibrium state, the KMS condition is a state-valued integral equation and furnishes a global principle; using it, ϕ_β can be computed directly (Moya (1); Sewell (2)). There is every reason to prepare the way for global techniques by a blend of global and local ideas; this text is written in this way.

2

The ideal Fermi gas

2.1. Introduction

THE IDEAL Fermi gas is the exemplar for systems of large numbers of particles whose spins are odd half-integers. It is an oversimplified model in that the constituent particles of this model do not interact with one another, which is indicated by the word 'ideal' in the name of the model. In conjunction with the Pauli exclusion principle, this results in the absence of collective phenomena and only one pure thermodynamic equilibrium phase is present at all temperatures and densities (Fermi (1); Pauli (1 - 3)).

Typical real fermion systems are the conduction electrons in a metal, nuclear matter, superconductors, and liquid helium III. One sees immediately that inter-particle interactions and strong collective phenomena are the rule and not the exception. Why then do we study the model? Even were the ideal Fermi gas devoid of immediate physical significance—and this is not the case—its study would be a preparation for the subsequent study of interacting systems. Everything is as mathematically favourable here as it can be, which is always a good starting point. It is not facetious to say that a theory which could not analyse this model could hardly be a useful one.

It is now well established that, in the normal state of metals, certain electrons, known as conduction electrons, move relatively freely throughout the metal as a whole. These conduction electrons are responsible for the following features of metals: electrical and thermal conductivities, specific heats, Hall effects, and the Richardson effect (Born [1]).

The particles in the ideal Fermi gas model correspond to the conduction electrons in a metal and give good agreement with the observed behaviour of metals. This aspect of the model is considered in great detail in many texts emphasizing the physical aspects of statistical mechanics, e.g. Born [1], Huang [1], Landau and Lifschitz [1], and Schrödinger [1].

The reader will notice the absence of any discussion of 'most probable values' in the text, in contrast to what occurs in most statistical mechanics texts (cf. Schrödinger [1]; Ehrenfest and Ehrenfest [1]). The reason is that such a discussion relates to the physical applicability of the grand canonical Gibbs states $\{\phi_\beta^{(n)}\}$ and their limit, the global Gibbs state $\phi_\beta \in \mathfrak{S}(\mathscr{A})$. We assume here that the $\{\phi_\beta^{(n)}\}$ are indeed the proper states to compute. Given this assumption, the only question which arises is whether or not they have been computed correctly; the familiar form of the result gives us confidence that we have done so.

As far as we know, the first paper on the ideal Fermi gas putting it into perspective *vis-à-vis* the representations of the canonical anticommutation relations is due to Araki and Wyss (1). Later work relating to this model in its algebraic framework as such concerns its relation to factorial W^*-algebras (cf. Sakai [1], §4.4 for a discussion and references pertaining to this) and to quasi-free states (Manuceau, Rocca, and Testard (1); Manuceau and Verbeure (2); Powers (1,2); Powers and Størmer (1); Rocca, Sirugue, and Testard (1)).

2.2. Configuration space

This model consists of spinless† fermions ('electrons') which move in $\mathbb{R}^3$, hence $\Gamma = \mathbb{R}^3$. The local regions are taken to be those bounded open subsets of $\mathbb{R}^3$ which are of piecewise smooth boundary. Certain pathologies are prevented by the further restriction that each local region is *star-shaped* with respect to at least one interior point (Choquet [1], Vol. I, p.346).

For this family of local regions we shall customarily write $\mathscr{L} = \{V\}$ as in the introduction; the convenient notation $\mathscr{L}' = \mathscr{L} \cup \{\mathbb{R}^3\}$ will also be adopted.

Whenever we write $V_n \in \mathscr{L}$ it is to indicate the cube of edge nL centered at the origin:

$$V_n = \{x \in \mathbb{R}^3 : -nL/2 \leqslant x_i \leqslant +nL/2; L > 0; i = 1, 2, 3\}; \qquad (2.1)$$

the family of such cubes is denoted $\mathscr{M}_L = \{V_n : n \in \mathbb{N}\}$. Note that given any $V \in \mathscr{L}$ there is an integer n_0 such that $V \subset V_m$ for all $m \geqslant n_0$. Accordingly $\mathscr{M}_L$ is seen to be absorbing (for $\mathscr{L}$). Rather more general absorbing sets can be considered, but $\mathscr{M}_L$ will prove sufficient for our purposes. Two further properties of $\mathscr{M}_L$ to note are that it is ordered by set-theoretical inclusion, and it covers $\mathbb{R}^3$. Thus $\mathscr{M}_L$ fulfils the general requirements set forth in Chapter 1. As L is fixed and more or less unimportant until §2.8, we shall abbreviate $\mathscr{M}_L \equiv \mathscr{M}$ in what follows.

Were we to compute the thermodynamic Gibbs state for general sequences of volumes, matters would be technically more difficult than we shall find. What we intend to do is to compute the local Gibbs states only for the cubes V_n and only with periodic boundary conditions; and then to take the thermodynamic limit for this case. Araki and Wyss (1) have done the more general fermion case, and Lewis and Pulé (1) (Lewis (1)) the more general boson case. Our feeling is that the global equilibrium state ought to be found by strictly global methods, such as the KMS condition, translation invariance, gauge invariance, and prescribed global density. (This may be enough to specify a state as already mentioned; cf. Moya (1) and Sewell (2)). One may disagree with our point of view and wish to

† See §3.1 for a discussion of spin.

see a computation of the thermodynamic limit of a more general family of local states. For this the reader will have to consult the above references for the Fermi and Bose gases. For this reason—in addition to the ease of computation—we shall restrict our attention to cubes and periodic boundary conditions in what follows.

2.3. Fock-Cook space

One-electron wave functions are the elements of the Hilbert space $\mathbb{L}^2(V)$ of square integrable functions from V to $\mathbb{C}$ for each $V \in \mathscr{L}'$. More precisely, we identify functions which are equal almost everywhere; here and hereafter this abuse of terminology will be assumed (Choquet [1]; Riesz and Sz-Nagy [1]).

By $\mathbb{L}^2_{\mathrm{A}}(V^n)$ we mean the Hilbert space of square integrable antisymmetric functions (classes of functions actually, see above) of n variables, from V^n to $\mathbb{C}$; the square of the norm of any $f^{(n)} \in \mathbb{L}^2_{\mathrm{A}}(V^n)$ is given by the integral

$$\|f^{(n)}\|^2 = \int_{V^n} |f^{(n)}(\mathbf{x}_1, \ldots, \mathbf{x}_n)|^2 \; d\mathbf{x}_1 \ldots d\mathbf{x}_n. \tag{2.2}$$

Elements of $\mathbb{L}^2_{\mathrm{A}}(V^n)$ are the n-electron wave functions. The space of wave functions for any number of electrons is known as *Fock-Cook space* (Cook (1); Fock (1)). It is the completed Hilbert direct sum

$$\mathscr{H}_F(V) = \bigoplus_{n=0}^{\infty} \mathbb{L}^2_{\mathrm{A}}(V^n), \tag{2.3a}$$

for every $V \in \mathscr{L}'$. By convention, the $n=0$ term is taken to be $\mathbb{C}$. See Gel'fand and Vilenken [1] and Guichardet [1 - 3] for more about this concept.† The symbol $\mathscr{H}_{\mathrm{F}}$ here is the $\mathscr{H}$ of Chapter 1 with the F subscript standing for *fermion.*

The notation for a typical vector $\Phi \in \mathscr{H}_{\mathrm{F}}(V)$ is $\Phi = \oplus \Phi^{(n)}$, where $\Phi^{(n)}$ is an element of $\mathbb{L}^2_{\mathrm{A}}(V^n)$ and the infinite series defining the square of the norm is assumed to converge:

$$\|\Phi\|^2 = \sum_{n=0}^{\infty} \|\Phi^{(n)}\|^2 < \infty. \tag{2.3b}$$

Going the other way, the operator projecting $\mathbb{L}^2_{\mathrm{A}}(V^n)$ out of $\mathscr{H}_{\mathrm{F}}(V)$ shall be denoted $P_V^{(n)}$. Thus

$$\begin{aligned} P_V^{(n)} &: \mathscr{H}_{\mathrm{F}}(V) \to \mathbb{L}^2_{\mathrm{A}}(V^n), \\ P_V^{(n)} &: \qquad \Phi \mapsto \Phi^{(n)}. \end{aligned} \tag{2.4}$$

The Fock-Cook spaces for different $V \in \mathscr{L}$ are *isotonic*, by which we mean that $\mathscr{H}_{\mathrm{F}}(V)$ is identifiable with a subspace of every $\mathscr{H}_{\mathrm{F}}(W)$ for which $V \subset W \in \mathscr{L}$. In fact, if $U, V \in \mathscr{L}$ are disjoint: $U \cap V = \emptyset$, then (Emch [1]; Guichardet [1 - 3]; Ruelle [1])

† Analogous structures occur in differential geometry, cf. Auslander [1] and Sternberg [1].

$$\mathscr{H}_{\mathrm{F}}(U \cup V) \simeq \mathscr{H}_{\mathrm{F}}(U) \otimes \mathscr{H}_{\mathrm{F}}(V). \tag{2.5}$$

If we wish to identify $\mathscr{H}_{\mathrm{F}}(U)$ with a subspace of $\mathscr{H}_{\mathrm{F}}(U \cup V)$ as in Chapter 1, we must distinguish a neutral unit vector in $\mathscr{H}_{\mathrm{F}}(V)$ in eqn (2.5). Conventionally, this vector is the so-called Fock-Cook vacuum vector, and is studied in detail in §2.5 (p. 28). It suffices here to know that there is some chosen unit vector $\Omega_V \in \mathscr{H}_{\mathrm{F}}(V)$ so that $\Phi \in \mathscr{H}_{\mathrm{F}}(U)$ is indentified with $\Phi \otimes \Omega_V$ in accordance with (2.5). For U and $V \in \mathscr{M}$ this will define the requisite injective mapping leading to eqn (2.9) below.

Because Fermi field operators satisfy canonical anticommutation relations (CAR), it will prove very convenient to define the *even* part of $\mathscr{H}_{\mathrm{F}}(V)$, $V \in \mathscr{L}'$, namely,

$$\mathscr{H}^{\mathrm{e}}_{\mathrm{F}}(V) = \bigoplus_{r=0}^{\infty} \mathbf{L}^2_{\mathrm{A}}(V^{2r}). \tag{2.6}$$

In view of eqns (2.4) and (2.6), the operator projecting out this subspace is the direct sum of the $P_V^{(n)}$ for even n:

$$P^e_V = \bigoplus_{r=0}^{\infty} P_V^{(2r)}, \tag{2.7}$$

so that we may write

$$P^e_V\,[\mathscr{H}_{\mathrm{F}}(V)] = \mathscr{H}^{\mathrm{e}}_{\mathrm{F}}(V). \tag{2.8}$$

The family $\{\mathscr{H}^{\mathrm{e}}_{\mathrm{F}}(V) : V \in \mathscr{L}\}$ of Hilbert spaces are also isotonic, eqns (2.5)-(2.8) leading to the analogue of eqn (2.5),

$$\mathscr{H}^{\mathrm{e}}_{\mathrm{F}}(U \cup V) \simeq \mathscr{H}^{\mathrm{e}}_{\mathrm{F}}(U) \otimes \mathscr{H}^{\mathrm{e}}_{\mathrm{F}}(V) \tag{2.5a}$$

for arbitrary disjoint local regions.

Hereafter, we shall let $\mathscr{H}^{\#}_{\mathrm{F}}$ stand for either $\mathscr{H}_{\mathrm{F}}$ or $\mathscr{H}^{\mathrm{e}}_{\mathrm{F}}$ indifferently. Then the absorbing property of the family $\mathscr{M} = \{V_n\}$ of cubes (eqn (2.1)) implies that $\mathscr{H}_{\mathrm{F}}(\mathbf{R}^3)$ is the Hilbert space inductive limit of the $\mathscr{H}_{\mathrm{F}}(V_n)$. For eqns (2.5) and (2.5a) imply the existence of a basis-independent injective mapping (cf. eqn (1.21)) $i^{\#}_{nm}$ of $\mathscr{H}^{\#}_{\mathrm{F}}(V_n)$ into $\mathscr{H}^{\#}_{\mathrm{F}}(V_m)$ for all $m \geqslant n$ (by using the Fock-Cook vacuum, cf. §2.5 (p.28). Then

$$\mathscr{H}_{\mathrm{F}}(\mathbf{R}^3) = \varinjlim \{i_{nm}\,[\mathscr{H}_{\mathrm{F}}(V_n)]\,; m \geqslant n; n, m \in \mathbf{N}\}; \tag{2.9a}$$

or simply

$$\mathscr{H}_{\mathrm{F}}(\mathbf{R}^3) = \varinjlim \mathscr{H}_{\mathrm{F}}(V_n) \tag{2.9b}$$

when no confusion is likely; this proves eqn (1.1) for this model. Hereafter, $\mathscr{H}_{\mathrm{F}}(V_n)$ will be written as $\mathscr{H}_n$, $\mathscr{H}_{\mathrm{F}}(V)$ as $\mathscr{H}(V)$, and $\mathscr{H}_{\mathrm{F}}(\mathbf{R}^3)$ as $\mathscr{H}$. These label omissions are not likely to cause any confusions within this chapter (Guichardet [1 - 3] discusses such structures).

2.4. The algebras

2.4.1. The quantum fields

The fermion annihilation and creation operators are defined everywhere on $\mathscr{H}(V)$ for $V \in \mathscr{L}'$ through the formulae† (cf. eqn (2.4))

$$P_V^{(n)}\,[a_V(f)\,\Phi]\,(\mathbf{y}_1,\ldots,\mathbf{y}_n) = (n+1)^{\frac{1}{2}} \int_V f(\mathbf{x}) P_V{}^{(n+1)}\,[\Phi]\,(\mathbf{x},\mathbf{y}_1,\ldots,\mathbf{y}_n)\,\mathrm{d}\mathbf{x} \tag{2.10a}$$

$$P_V^{(n)}\,[a_V^*(f)\,\Phi]\,(\mathbf{y}_1,\ldots,\mathbf{y}_n) = (n)^{-\frac{1}{2}} \sum_{j=1}^{n} (-1)^j P_V{}^{(n-1)}\,[\Phi]\,(\mathbf{y}_1,\ldots,\hat{\mathbf{y}}_j,\ldots,\mathbf{y}_n)\,f(\mathbf{y}_j) \tag{2.10b}$$

for every $f \in \mathbb{L}^2(V)$. The V-subscript on these field operators can usually be omitted, it being deducible from the context. Another notational convenience is to write $a^\#(f)$ for either $a(f)$ or $a^*(f)$ indifferently (Emch [1]; Guichardet [4]; Ruelle [1]; Streater and Wightman [1]).

These fermion field operators obey the CAR:

$$\{a^*(f), a(\bar{g})\} = (f,g)_V\,\mathbf{1}\ , \tag{2.11a}$$

$$\{a(f), a(g)\} = 0. \tag{2.11b}$$

The braces denote the anticommutator, and $(.,.)_V$ is the inner product for $\mathbb{L}^2(V)$.

$a^\#(f)$ is complex-linear from $\mathbb{L}^2(V)$ into $\mathscr{H}(V)$:
$a^*(f+g) = a^*(f) + a^*(g)$, $a^*(zf) = za^*(f)$, and $a^*(f) = [a(\bar{f})]^*$; z is any element of $\mathbb{C}$, and $\bar{f}$ is the complex conjugate of f, with $f, g \in \mathbb{L}^2(V)$, $V \in \mathscr{L}'$.

Let us now show that the field operators are bounded by finding their norms. To do this, we form the 'f-mode number operator'

$$N_V(f) = a_V^*(f)\,a_V(\bar{f}). \tag{2.12}$$

Using the CAR, it is clear that $N_V(f)$ is a projection operator on $\mathscr{H}(V)$, and using the property $\|A^*A\| = \|A\|^2$ satisfied by the C^*-norm, it is trivial to verify that $\|N_V(f)\|^2 = \|f\|^2$. By the same property of the norm, the definition of $N_V(f)$ then leads to the norm of $a(f)$, namely,

$$\|a^\#(f)\| = \|f\| \quad (f \in \mathbb{L}^2(V)). \tag{2.13}$$

2.4.2. The local algebras

The local Fermi field algebra for the region V, written $\mathscr{A}(V)$, is taken to be the C^*-algebra of bounded operators on Fock-Cook space $\mathscr{H}_{\mathrm{F}}(V) \equiv \mathscr{H}(V)$ generated by the fermion fields defined above. In an obvious notation, with the

† In eqn (2.10b), $\hat{y}_j$ means that the variable y_j is omitted.

angular braces indicating the algebraic span, i.e. polynomials in the indicated variables, this means that

$$\mathscr{A}(V) = \text{un. cl.} - \langle a_V(f) \; : \; \forall f \in \mathbf{L}^2(V) \rangle. \tag{2.14}$$

We shall also refer to $\mathscr{A}(V)$ as the CAR algebra for the region V.

Because of the anticommutation relations, the family of C^*-algebras $\{\mathscr{A}(V)\}$ is precluded from consideration as candidate for the family of local observable algebras. For $[\mathscr{A}(V), \mathscr{A}(W)]_- \neq 0$ for disjoint local regions: $V \cap W = \emptyset$. These CAR algebras will be seen to be most useful none the less.

As the local observable algebras we shall choose the C^*-algebra of bounded operators on the *even* Fock-Cook space generated by the fermion fields:

$$\mathscr{A}^{\mathrm{e}}(V) = \mathscr{A}(V) \mid \mathscr{H}^{\mathrm{e}}(V). \tag{2.15}$$

Hereafter, we shall write $\mathscr{A}^{\#}(V)$ for $\mathscr{A}(V)$ or $\mathscr{A}^{\mathrm{e}}(V)$ indifferently.

2.4.3. *Local gauge transformations*

It is clearly of interest to relate $\mathscr{A}(V)$ and $\mathscr{A}^{\mathrm{e}}(V)$ for some fixed $V \in \mathscr{L}$. Using the even projector P^{e}_V defined in eqns (2.7) and (2.8), we have the simple relation

$$\begin{aligned} \mathscr{A}^{\mathrm{e}}(V) &= P^{\mathrm{e}}_V \mathscr{A}(V) P^{\mathrm{e}}_V \\ &= \{P^{\mathrm{e}}_V A P^{\mathrm{e}}_V \; : \; A \in \mathscr{A}(V)\}. \end{aligned} \tag{2.16}$$

Equivalently, one may employ the number operator on Fock-Cook space to relate these algebras to each other. The number operator N_V is an unbounded self-adjoint operator whose domain is

$$\mathfrak{E}(V) = \left\{ \Phi \in \mathscr{H}_{\mathrm{F}}(V) \; : \; \sum_{n=0}^{\infty} n^2 \, \|\Phi^{(n)}\|^2 < \infty \right\}, \tag{2.17a}$$

which latter subset is dense in $\mathscr{H}_{\mathrm{F}}(V)$. (See Kato [1] for general matters pertaining to unbounded operators.) The formula for the operator is

$$N_V = \bigoplus_{n=0}^{\infty} n P^{(n)}_V. \tag{2.17b}$$

The number operator N_V generates the unitary group

$$\begin{aligned} U_V(\theta) &= \exp(\mathrm{i}\theta N_V) \\ &= \bigoplus_{n=0}^{\infty} \mathrm{e}^{\mathrm{i}n\theta} P^{(n)}_V \quad (-\pi \leqslant \theta < +\pi). \end{aligned} \tag{2.18}$$

Noting that $U_V(0)$ is the unit operator, the even projection operator is related to the number operator by the formula $P^{\mathrm{e}}_V = \frac{1}{2}[\mathbf{1}_V + U_V(-\pi)]$.

In eqn (2.17) we have a subsystem number operator N_V, related to the structure of Fock-Cook space through the $\{P^{(n)}_V \; : \; n \in \mathbf{N}\}$; in eqn (2.12) we have a mode-number operator $N_V(f)$ related to the quantum fields. Note that it follows from (2.12) that $N_V(f) \in \mathscr{A}^{\mathrm{e}}(V)$, as it is a second-degree polynomial in the fields.

The familiar connection between N_V and $N_V(f)$ has been rigorously demonstrated by Chaiken (1,2), Dell'Antonio and Doplicher (1), and Dell'Antonio, Doplicher, and Ruelle (1) for *any* orthonormal basis $\{e_p : p \in \mathbf{N}\}$ of $\mathbf{L}^2(V)$ for $V \in \mathscr{L}'$ (i.e. including $V \equiv \mathbf{R}^3$)

$$N_V = (\text{gn. str. } \mathscr{H}_F(V)) - \lim_{p\to\infty} \sum_{j=1}^{p} N_V(e_j). \tag{2.19a}$$

Generalized strong (gn. str.) limits are strong limits on the pertinent operator domains. Kato [1] discusses this, emphasizing the resolvent operator convergence.

Abbreviating the generalized strong limit indicator by a superscripted s on the summation symbol, this can be written in the more usual form

$$N_V = \sum_{p\in \mathbf{N}}^{(s)} a_V^*(e_p)\, a_V(\bar{e}_p). \tag{2.19b}$$

But then it is clear that $U_V(\theta) \in \mathscr{A}^{\#}(V)''$, and generates a symmetry of the system by similarity transform. If we write $T = \{-\pi \leqslant \theta < +\pi\}$ for the additive group of angles, then

$$\begin{aligned} \Gamma_V &: T \to \text{Aut}\,[\mathscr{A}^{\#}(V)],\\ \Gamma_V(\theta)A &= U_V(\theta)A\,U_V(-\theta) \end{aligned} \tag{2.20}$$

is the corresponding group of automorphisms.

In elementary one-particle quantum mechanics, gauge transformations of the first kind transform electron wave functions in accordance with the rule $\psi(\mathrm{x}) \to \exp(ie\lambda)\,\psi(\mathrm{x})$ with space-independent $\lambda \in \mathbf{R}$. By analogy, we interpret Γ_V as the automorphism group of gauge transformations of the first kind for the (sub)system associated with $V \in \mathscr{L}'$. In this connection, cf. Dell'Antonio (2), Gille and Manuceau (1), Manuceau (1), and Rocca and Sirugue (1).

The following additional remarks might be of interest.

(1) The subalgebra of gauge-invariant elements of $\mathscr{A}(V)$, those for which $\Gamma_V(\theta)A = A$ for every $\theta \in T$, is a proper subalgebra of $\mathscr{A}^{e}(V)$.

(2) The number operator $N(\mathbf{R}^3)$ for $\mathbf{R}^3$ exists and is self-adjoint on the subspace $\mathfrak{E}(\mathbf{R}^3) \subset \mathscr{H}_F(\mathbf{R}^3)$.

(3) The number operator N_{V_n} for $V \in \mathscr{L}$ is the operator which appears in the reduced Hamiltonian $\bar{H}_n = H_n - \mu_n N_n$ of (1.28).

2.4.4. *Local space translations*

We now consider space translations. We must give an explicit definition for the $S_V(\xi)$ ($\xi \in \mathbf{R}^3$) and check that the formulae of §1.7 hold.

For every $\Phi \in \mathscr{H}_F(V)$ and every vector $\xi \in \mathbf{R}^3$, the requisite isometry is defined to be

$$\begin{aligned} &S_V(\xi) : \mathscr{H}_F(V) \to \mathscr{H}_F(V+\xi);\\ &P_V^{(n)}[S_V(\xi)\Phi](\mathrm{x}_1,\ldots,\mathrm{x}_n) = [P_V^{(n)}\Phi](\mathrm{x}_1-\xi,\ldots,\mathrm{x}_n-\xi). \end{aligned} \tag{2.21}$$

The propriety of this choice depends in the final analysis upon the translational invariance of Lebesgue measure on $\mathbb{R}^3$ (Choquet [1]) and on the formula

$$\int_V |f(\mathrm{x}-\xi)|^2\,\mathrm{dx} = \int_{V+\xi} |f(\mathrm{x})|^2\,\mathrm{dx} \tag{2.22}$$

for, say, functions $f \in \mathscr{S}(\mathbb{R}^3)$ (cf. Gel'fand and Shilov [1,2], Gel'fand and Vilenkin [1], and Trèves [1] concerning spaces of distributions). It follows that $S_V(\xi)$ is an isometry of $\mathscr{H}_{\mathrm{F}}(V)$ onto $\mathscr{H}_{\mathrm{F}}(V+\xi)$, as $\mathbb{L}^2(V^n)$ has polynomials in n-fold product tensors (such as $f \otimes g \otimes \ldots \otimes h$) as a dense set.

The corresponding formula for the local subalgebras has already been given in §1.7:

$$\sigma_V(\xi) : \mathscr{A}^{\#}(V) \to \mathscr{A}^{\#}(V+\xi);$$
$$\sigma_V(\xi)A = S_V(\xi)A\,S_V(-\xi). \tag{2.23}$$

2.4.5. The quasilocal algebras

We shall now form the quasilocal algebras in the manner described in Chapter 1. As in eqn (1.22) we write $j^{\#}(V,W)$ for the injection of $\mathscr{A}^{\#}(V)$ into $\mathscr{A}^{\#}(W)$ when $V \subset W \in \mathscr{L}$; and $j^{\#}_{nm}$ as an abbreviation for $j^{\#}(V_n, V_m)$ $(m \geqslant n)$. Then the quasilocal algebras are the C^*-inductive limits (cf. (1.2))

$$\mathscr{A}^{\#} = \varinjlim \{ j_{nm}(\mathscr{A}^{\#}_n); n, m \in \mathbf{N}; m \geqslant n \}. \tag{2.24}$$

The systemic local algebras are the unions of the local subalgebras as in eqn (1.3):

$$\mathscr{A}^{\#}_L = \cup_{V \in \mathscr{L}}\, j^{\#}(V)\,[\mathscr{A}^{\#}(V)]. \tag{2.25}$$

The map $j^{\#}(V)$ injects $\mathscr{A}^{\#}(V)$ into $\mathscr{A}^{\#}$, making the necessary identifications explicit. By the theorem quoted in §1.2, eqn (1.4) holds for this model (cf. eqn (1.4)); accordingly

$$\mathscr{A}^{\#} = \text{un. cl.-}\,(\mathscr{A}^{\#}_{\mathrm{L}}). \tag{2.26}$$

The first point to note is that the $\{\mathscr{A}^{\mathrm{e}}(V) : V \in \mathscr{L}\}$ do, but the $\{\mathscr{A}(V) : V \in \mathscr{L}\}$ do not obey the local commutativity constraints

$$[\mathscr{A}^{\mathrm{e}}(V), \mathscr{A}^{\mathrm{e}}(W)]_- = 0, V \cap W = \emptyset. \tag{2.27}$$

The second point to note concerns the fields generating $\mathscr{A}^{\#}$. From (2.25) and (2.26) it follows that $a^{\#}(f) \in \mathscr{A}$ for every $f \in \mathbb{L}^2(\mathbb{R}^3)$ of compact support. But even more is true; because $f \to a^{\#}(f)$ is bounded, $a^{\#}(f) \in \mathscr{A}$ for every $f \in \mathbb{L}^2(\mathbb{R}^3)$ with no restrictions.† (This last property is not true for the boson fields.) In view of this, let us write

$$\mathscr{A} = \text{un. cl.-} \langle a^{\#}(f) : \forall f \in \mathbb{L}^2(\mathbb{R}^3) \rangle, \tag{2.24a}$$

where, recall, $\langle\ .\ \rangle$ stands for the algebraic span.

† There is an implicit extension by continuity here.

2.5. The Fock-Cook state

There is a certain vector Ω_V in $\mathscr{H}_{\mathrm{F}}(V)$ (for $V \in \mathscr{L}'$) which is especially useful and important. Known as the Fock-Cook vacuum vector, it is defined by

$$\Omega_V = 1 \oplus 0 \oplus 0 \oplus \ldots \quad (V \in \mathscr{L}'). \tag{2.28}$$

Its associated vector state $\omega_V^{(0)} \in \mathfrak{S}\,[\mathscr{A}^{\#}(V)]$ is known as the Fock-Cook state (Cook (1); Fock (1)):

$$\omega_V^{(0)}(A) = (\Omega_V, A\,\Omega_V)_{\mathscr{H}_{\mathrm{F}}(V)} \qquad (A \in \mathscr{A}^{\#}(V)). \tag{2.29}$$

Note that we shall not distinguish between the Fock-Cook state on $\mathscr{A}(V)$ and its restriction to $\mathscr{A}^{\mathrm{e}}(V)$.

If we apply the defining action eqn (2.10a) of a_V to the vacuum vector Ω_V (eqn (2.28)), we find that a_V annihilates Ω_V:

$$a_V(f)\,\Omega_V = 0 \quad (\forall f \in \mathbb{L}^2(V)); \tag{2.30}$$

this equation shall be called the Fock-Cook condition.

In the same way we can compute $\omega_V^{(0)}$ explicitly, since it is sufficient to consider monomial arguments of the form

$$\mathscr{X}_{nm} \equiv a_V(\bar{f}_1) \ldots a_V(\bar{f}_n)\, a_V^*(g_1) \ldots a_V^*(g_m); \tag{2.31}$$

there is no loss of generality in having taken the operators ordered in this way, for, by using the CAR and the Fock-Cook condition, any unordered monomial can be cast into the canonical order.

Then an inductive computation yields

$$\omega_V^{(0)}(\mathscr{X}_{nm}) = \delta_{nm} \det |(f_j, g_k)|, \tag{2.32a}$$

e.g.

$$\omega_V^{(0)}\,[a_V(\bar{f})\,a_V^*(g)] = (f,g)_V, \tag{2.32b}$$

where $(.\,,.)_V$ is the $\mathbb{L}^2(V)$ inner product.

Upon combining (2.32a) and (2.32b), $\omega_V^{(0)}$ takes the interesting form of being a determinant of its own two-point functions:

$$\omega_V^{(0)}(\mathscr{X}_{nm}) = \delta_{nm} \det |\,\omega_V^{(0)}\,[a_V(\bar{f}_j)\,a_V^*(g_k)]\,|. \tag{2.32c}$$

This form is characteristic of a class of states on $\mathscr{A}^{\#}(V)$ known as quasi-free states: $\psi \in \mathfrak{S}\,[\mathscr{A}^{\#}(V)]$ is *quasi free* iff ψ is of the form:

$$\psi(\mathscr{X}_{nm}) = \delta_{nm} \det |\,\psi\,[a_V(\bar{f}_j)\,a_V^*(g_k)]\,|.$$

The terminology is quite apt, as such states have certain very characteristic properties in common with $\omega_V^{(0)}$; quasi-free states for $\mathscr{A}^{\#}(V)$ have been extensively analysed in a series of papers; the list of references given here includes the analogous analysis for bosons and the KMS condition: Manuceau (1), Manuceau, Rocca, and Sirugue (1), Manuceau and Verbeure (1), Powers (1,2), Powers and

Størmer (1), and Rocca, Sirugue, and Testard (1,2).

We now wish to make the following observation. Recall that $\mathscr{A}_n^{\#}$ is identified with a subalgebra of $\mathscr{A}^{\#}$ under the mapping $A^{(n)} \mapsto A^{(n)} \otimes \mathbb{1}_n$, where $\mathbb{1}_n$ is the identity operator on $\mathscr{H}_{\mathrm{F}}(\mathbb{R}^3 \backslash V_n)$. To do the same for states, we must have a distinguished neutral state on $\mathscr{A}^{\#}(\mathbb{R}^3 \backslash V_n)$, just as $\mathbb{1}_n$ is a neutral operator. The canonical candidate is clearly the Fock-Cook state $\omega^{(0)}(\mathbb{R}^3 \backslash V_n)$. That is, for any state $\psi_n \in \mathfrak{S}(\mathscr{A}_n^{\#})$, the mapping

$$\psi_n \mapsto \psi_n \otimes \omega^{(0)}(R^3 \backslash V_n) \equiv \tilde{\psi}_n \tag{2.33}$$

identifies ψ_n with a state $\tilde{\psi}_n$ on $\mathscr{A}^{\#}$. Of course, one could choose a different product state to be the neutral state. One would then modify eqn (2.33) accordingly, but we have no need to do so in this work.

The use of this device is that a sequence of states $\{\psi_n \in \mathfrak{S}(\mathscr{A}_n^{\#}) \; : \; n \in \mathbb{N}\}$ can be examined for eonvergence (or no) to a state on $\mathscr{A}^{\#}$ by examining the $\{\tilde{\psi}_n \in \mathfrak{S}(\mathscr{A}^{\#}) \; : \; n \in \mathbb{N}\}$ which are all states on the same algebra.

In this sense the Fock-Cook states themselves converge in the weak*-topology to the Fock-Cook state on $\mathscr{A}^{\#}$. For it follows from eqns (2.32) and (2.33) that

$$\lim_{n\to\infty} |\tilde{\omega}_n^{(0)}(A) - \omega^{(0)}(A)| = 0 \quad (\forall A \in \mathscr{A}_{\mathrm{L}}^{\#}). \tag{2.34a}$$

This limit is written in the abbreviated form

$$\text{weak*-}\lim_{n\to\infty} \tilde{\omega}_n^{(0)} = \omega^{(0)}, \tag{2.34b}$$

since the convergence extends from $\mathscr{A}_{\mathrm{L}}^{\#}$ to $\mathscr{A}^{\#}$. We have set

$$\omega^{(0)} \equiv \omega_{\mathbb{R}^3}^{(0)} \text{ and } \omega_n^{(0)} \equiv \omega_{V_n}^{(0)} \qquad \text{for brevity.}$$

Some last remarks about $\omega^{(0)}$.

(1) From its determinantal form, it follows that $\omega^{(0)}(B) = 0$ for every $B \in \mathscr{A} \backslash \mathscr{A}^{\mathrm{e}}$; we say that $\omega^{(0)}$ is an *even state.*

(2) The GNS construction associated with $\omega^{(0)}$ leads to $\omega^{(0)} \sim [\mathscr{H}_{\mathrm{F}}, \mathrm{id}, \Omega_{\mathbb{R}^3}]$, where $\mathrm{id}(A) = A$ is the identity mapping.

(3) $\omega^{(0)}$ is space translationally invariant, as $\|S(\xi)f\|^2 = \|f\|^2$ for every $\xi \in \mathbb{R}^3$, $f \in \mathbb{L}^2(\mathbb{R}^3)$.

2.6. The spin algebra

It has been known for a long time that there is a close connection between fermion fields and spin-$\frac{1}{2}$ operators (Jordan and Wigner (1)). One may even reduce the study of the CAR algebras to a study of spin algebras, as has been done by Guichardet [4] (cf. Emch [1], and references there). Our primary interest in this connection is computational; we shall use this connection to compute the grand partition function for the ideal Fermi gas. In addition we shall establish notation which will prove useful in our discussion of spin lattices.

First we shall construct the spin algebra, and then connect it to the CAR algebras. Let $\mathbb{C}^2$ be the Hilbert space of complex pairs and $\mathbb{B}(\mathbb{C}^2)$ the W^*-algebra of all bounded operators on it.

For each whole number $p \in \mathbb{N}$ let $\mathfrak{B}_p$ be a copy of $\mathbb{B}(\mathbb{C}^2)$. That is, there exists a C^*-isomorphism

$$\gamma_p : \mathbb{B}(\mathbb{C}^2) \to \mathfrak{B}_p. \tag{2.35}$$

It is well known that in general there is no unique topological tensor product of two C^*-algebras; but there is in all cases a minimal cross norm, and it is canonical to choose the completion in this topology as *the* C^*-tensor product. If $\mathfrak{A}_1$ and $\mathfrak{A}_2$ are arbitrary C^*-algebras, we shall write $\mathfrak{A}_1 \otimes \mathfrak{A}_2$ for this completion. No confusion is likely, as no other C^*-tensor product is ever used in this text. See Grothendieck [1]; Guichardet [1 - 3], and Sakai [1] for further references; note that the symbol $\mathfrak{A}_1 \otimes_{\alpha_0} \mathfrak{A}_2$ is used by Sakai for this completion.

All this is very general, but often certain special results can be used. This is the case here. For $\mathfrak{B}_p$ is finite-dimensional: it is generated by four linearly-independent 2×2 matrices. As such, the algebraic tensor product $\mathfrak{B}_p \odot \mathfrak{B}_q$ is *already complete* and no problem arises. Hence it is the *algebraic tensor product* which appears for the spin algebras (Auslander [1]; Sternberg [1]).

With this notation we shall write

$$\mathfrak{B}[n] = \bigotimes_{p=1}^{n} \mathfrak{B}_p \quad (n \in \mathbb{N}). \tag{2.36}$$

for the C^*-tensor product (which is here the algebraic tensor product). If $m > n$, there is a canonical injection of $\mathfrak{B}[n]$ into $\mathfrak{B}[m]$, namely,

$$\begin{aligned} \varphi_{nm} &: \mathfrak{B}[n] \to \mathfrak{B}[m]; \\ \varphi_{mn}(A) &= A \otimes \mathbb{1}_{m\backslash n} \quad (m > n). \end{aligned} \tag{2.37}$$

It is clear from this that these C^*-algebras form an inductive family; letting $\mathfrak{B}$ denote the inductive limit C^*-algebra, we write

$$\mathfrak{B} = \varinjlim \{\varphi_{mn}(\mathfrak{B}[n]) : m,n \in \mathbb{N}; m > n\}. \tag{2.38}$$

Note that $\mathfrak{B}$ is not a finite-dimensional algebra although it enjoys many of their properties.

There are a number of identification and injection mappings which we shall make explicit here: $\mathfrak{B}_p$ into $\mathfrak{B}[n]$, $\mathfrak{B}[n]$ into $\mathfrak{B}$, etc. All of these are formed the same way: one places an identity operator in the neutral modes. Thus in an obvious notation

$$\begin{aligned} \varphi_n &: \mathfrak{B}[n] \to \mathfrak{B} \\ \varphi_n(A) &= A \otimes \mathbb{1}_{\mathbb{N}\backslash n} \qquad (A \in \mathfrak{B}[n]), \end{aligned} \tag{2.39}$$

$$\begin{aligned} \psi_p^n &: \mathfrak{B}_p \to \mathfrak{B}[n] \qquad (p \leqslant n) \\ \psi_p^n(A) &= \bigotimes_1^{p-1} \mathbb{1} \otimes A \bigotimes_{p+1}^{n} \mathbb{1} \quad (A \in \mathfrak{B}_p). \end{aligned} \tag{2.40}$$

Of course, $\varphi_n \circ \psi_n^p : \mathfrak{B}_p \to \mathfrak{B}$ maps $\mathfrak{B}_p$ continuously into $\mathfrak{B}$; and $\varphi_n \circ \psi_p^n \circ \gamma_p : \mathbb{B}(\mathbb{C}^2) \to \mathfrak{B}$ maps $\mathbb{B}(\mathbb{C}^2)$ continuously into $\mathfrak{B}$ with only the pth mode non-neutral, i.e. here the n is a 'dummy variable'.

The algebra $\mathbb{B}(\mathbb{C}^2)$ is generated by the Pauli spin matrices $\{\sigma^{(\alpha)} : \alpha = 1,\ldots,4\}$, with $\sigma^{(4)}$ written for the 2×2 identity matrix; we shall also set $\sigma^{(1)} \pm i\sigma^{(2)}$ equal to $\sigma^{(\pm)}$.

As these spin operators generate $\mathbb{B}(\mathbb{C}^2)$, and the algebras $\mathfrak{B}[n]$, $\mathfrak{B}$ are formed from $\mathbb{B}(\mathbb{C}^2)$, there are generating spin operators for these latter algebras, formed from the Pauli matrices. We shall write

$$J_n^{(\alpha)}(m) = \psi_n^m \circ \gamma_n[\sigma^{(\alpha)}] \in \mathfrak{B}[m] \quad (m \geqslant n) \tag{2.41}$$

and simply

$$J_n^{(\alpha)} = \varphi_m[J_n^{(\alpha)}(m)] \in \mathfrak{B} \tag{2.42}$$

for any $m \geqslant n$. Thus $\{J_n^{(\alpha)}(m) : n = 1,\ldots,m; \alpha = 1,\ldots,4\}$ (resp. $\{J_n^{(\alpha)} : n \in \mathbb{N}; \alpha = 1,\ldots,4\}$) form a generating set for $\mathfrak{B}[m]$ (resp. $\mathfrak{B}$). These generating sets satisfy relations, which follow from

$$\sigma^{(\alpha)}\sigma^{(\beta)} = \delta_{\alpha\beta} + i\sum_{\gamma=1}^{3} \epsilon_{\alpha\beta\gamma}\,\sigma^{(\gamma)} \qquad (\alpha,\beta = 1,2,3).$$

These are

$$[J_p^{(\alpha)}, J_q^{(\beta)}]_- = 2i\,\delta_{pq}\sum_{\gamma=1}^{3} \epsilon_{\alpha\beta\gamma}\,J_q^{(\gamma)} \qquad (\alpha,\beta = 1,2,3), \tag{2.43}$$

and similar ones for the $\mathfrak{B}[n]$ generators.

The relation between the Fermi local subalgebra $\mathscr{A}(V)$ and $\mathfrak{B}$ is not *natural*: it depends upon the choice of an orthonormal basis for $\mathbb{L}^2(V)$. It will be sufficient to consider $V_n \in \mathscr{M}$, the cube of edge nL as defined in eqn (2.1). We shall also use the particular basis of $\mathbb{L}^2(V_n)$ adapted to toroidal boundary conditions

$$\{e_\kappa : \kappa \in \widetilde{V}_n; e_\kappa(\xi) = (nL)^{-3/2}\exp(i\kappa \cdot \xi)\}, \tag{2.44a}$$

where $\xi \in V_n$ and where $\widetilde{V}_n$ is the 'toroidal dual' of V_n:

$$\widetilde{V}_n = \frac{2\pi}{nL}\mathbb{Z}^3. \tag{2.44b}$$

As we shall see, this set is very important in our computations.

It is very useful to order the vectors in $\widetilde{V}_n$ in some way, so as to be able to label the e_κ by a strictly positive integer. As the cardinality of $\mathbb{Z}$ and $\mathbb{Z}^3$ are the same, this is possible; so let $\eta_n : \widetilde{V}_n \to \mathbb{N}$ be some bijection, and write†

$$a_p^{*(n)} = a_{V_n}^*(e_\kappa), \quad (\eta_n(\kappa) = p) \tag{2.45}$$

for the Fermi fields in $\mathscr{A}_n$.

† The Cantor process (Gamow [1]) will do. One proceeds inductively, ordering all triples (l,m,n), whose sum $l+m+n=N$ is fixed; there are $\frac{1}{2}(N+2)(N+1)$ of them. This ordering is $(l,m,n) < (l',m',n')$ if $l < l'$; if $l = l'$, then $m < m'$; if $m = m'$, then $n < n'$. Then those for which $l+m+n=N+1$, etc.

So defined, the set $\{a_p^{\#(n)} : p \in \mathbf{N}\}$ forms a family of independent Fermi oscillators in $\mathscr{A}_n$; they satisfy the CAR in the form

$$\{a_p^{*(n)}, a_q^{(n)}\} = \delta_{pq}, \tag{2.46}$$

other anticommutators vanishing. (This is why they are called independent.)

Corresponding to this choice of orthonormal basis and choice of bijection η_n we define the following connection between $\mathscr{A}_n$ and $\mathfrak{B}$:

$$\begin{aligned} \pi_n &: \mathscr{A}_n \to \mathfrak{B}, \\ \pi_n [a_p^{*(n)}] &= \prod_{q=1}^{p-1} J_q^{(3)} \cdot J_p^{(-)}, \\ \pi_n [a_p^{(n)}] &= \prod_{q=1}^{p-1} J_q^{(3)} \cdot J^{(+)}. \end{aligned} \tag{2.47}$$

There is no label indicating V_n on these spin operators, but we trust that, as V_n remains fixed for the time being, no confusion will result.

Let us note that eqn (2.47) will define a C^*-isomorphism $\pi_n : \mathscr{A}_n \to \mathfrak{B}$ if it exists. This point is cleared up by noting that π_n is bounded, as $\| \pi_n [a_p^{\#(n)}] \| = 1$. Further, $\pi_n(0) = 0$ and $\pi_n(A) = 0$ implies $A = 0$ from the very nature of $\mathscr{A}_n$ and $\mathfrak{B}$. For example, $\pi_n [a_p^{*(n)} a_p^{*(n)}] = J_p^{(-)} J_p^{(-)} = 0$, as it should, etc. Consequently, (2.47) is well defined.

An algebra is *simple* if it has no proper closed two-sided ideals. If a C^*-algebra $\mathfrak{A}$ is the C^*-inductive limit of a family of simple C^*-algebras $(\mathfrak{A}_n)_{n \in \mathbf{N}}$, say, then $\mathfrak{A}$ is simple (Sakai [1], proposition 1.23.8). The even CAR is simple, as was shown by Doplicher and Powers (1).

2.7. Time translations

2.7.1. Local time translations

An ideal gas is one for which the systemic dynamics follows from the one-particle evolution corresponding to the Laplacian. Having defined the orthonormal basis eqn (2.44) adapted to toroidal boundary conditions, it is trivial to define the corresponding self-adjoint one-particle Hamiltonian extended from the Laplacian by means of this basis, namely,

$$\bar{h}_n = \sum_{\kappa \in \widetilde{V}_n} \epsilon_n(\kappa) e_\kappa \otimes e_\kappa^* \tag{2.48}$$

defining $\bar{h}_n$ through its spectral decomposition†, where its eigenvalues are

$$\epsilon_n(\kappa) = \frac{\hbar^2 |\kappa|^2}{2m} - \mu_n;$$

recall that $\widetilde{V}_n = \frac{2\pi}{nL}\mathbf{Z}^3$ is the toroidal dual of V_n (eqn (2.44)).

† The expression $e_\kappa \otimes e_\kappa^*$ acts on any $f \in \mathbf{L}^2(V_n)$ as $e_\kappa \otimes e_\kappa^*(f) = [e_\kappa^*(f)] e_\kappa = \langle f, e_\kappa \rangle e_\kappa$, and so corresponds to Dirac's symbol $|\kappa\rangle\langle\kappa|$. The vector e_κ^* is the dual basis vector to e_κ, but as $\mathbf{L}^2(V)$ is self-dual through the inner product, the indicated result occurs.

The domain of h_n can be seen directly from its spectral resolution eqn (2.48):

$$\mathbb{D}(\bar{h}_n) = \left\{ f = \sum_{\kappa \in \widetilde{V}_n} (nL)^{-3/2} \widetilde{f}_\kappa e_\kappa : \sum_{\kappa \in \widetilde{V}_n} |\widetilde{f}_\kappa|^2 \epsilon_n(\kappa)^2 < \infty \right\}, \quad (2.49)$$

where the Fourier components are defined by

$$\widetilde{f}_\kappa = \int_{V_n} f(\xi) \exp(\mathrm{i}\kappa \cdot \xi) \, \mathrm{d}\xi \qquad \text{(in the mean)}$$

$$= (nL)^{3/2} \langle f, e_\kappa \rangle. \quad (2.50)$$

The connection with the Laplacian is this: let $\dot{h}_n$ be the formal elliptic differential operator $-\frac{\hbar^2}{2m} \triangle_n - \mu_n$ on V_n. If $g \in \mathbb{D}(\bar{h}_n)$ is twice continuously differentiable and satisfies $g(\xi + \mathrm{a}) = g(\xi)$ for every $\xi \in V_n$, $\mathrm{a} \in nL\mathbb{Z}^3$, then $\dot{h}_n g = \bar{h}_n(g)$ (Kato [1]).

In our usual notation, $\bar{h}_n$ generates the one-parameter unitary group (resp. semi-group)

$$u_n^{(1)}(t) = \exp(\mathrm{i}t\,\bar{h}_n) \quad (t \in \mathbb{R}) \quad (2.51a)$$

resp.

$$\sigma_n^{(1)}(\beta) = \exp(-\beta \bar{h}_n) \quad (\beta \in \mathbb{R}^+)); \quad (2.51b)$$

the superscripted unity signifies that these are one-particle operators on $\mathbb{L}^2(V_n)$ and not operators on $\mathscr{H}_n$, the Fock-Cook Hilbert space. For the local time translations for the subsystem we proceed as follows. The pertinent automorphism group is

$$\tau^{(n)} : \mathbb{R} \to \mathrm{Aut}(\mathscr{A}_n^{\#});$$

$$\tau_t^{(n)} [a_n^{\#}(f)] = a_n^{\#} [u_n^{(1)}(t) f] \quad (\forall f \in \mathbb{L}^2(V_n)), \quad (2.52)$$

where $a_n^{\#}$ is the Fermi field in $\mathscr{A}_n$. That this defines an automorphism of $\mathscr{A}_n$ follows from the fact that $u_n^{(1)}(t) f \in \mathbb{L}^2(V_n)$ when f is. That this leads to an automorphism of $\mathscr{A}_n^e$ follows because $\tau^{(n)}(\mathbb{R})$ commutes with the gauge automorphism group $\Gamma_n(T)$ and hence with $P_{V_n}^e$, the even projection operator.

The expression

$$a_n^{\#} [u_n^{(1)}(t) f] = U_t^{(n)} a_n^{\#}(f)\, U_{-t}^{(n)} \quad (2.53)$$

serves to define the strongly-continuous one-parameter unitary group $U^{(n)}(\mathbb{R})$ which implements $\tau^{(n)}(\mathbb{R})$. The full Hamiltonian $\bar{H}_n$ generates this group:

$$U_t^{(n)} = \exp(\mathrm{i}t\bar{H}_n), \quad (2.54)$$

and it is a standard calculation to verify that, in terms of the quantum fields, $\bar{H}_n$ is given by

$$\bar{H}_n = \sum_{\kappa \in V_n}^{(s)} \epsilon_n(\kappa)\, a_n^*(e_\kappa)\, a_n(\bar{e}_\kappa), \quad (2.55)$$

where the sum is meant as a limit of the partial sums in the generalized strong sense. Furthermore, each of these partial sums is a polynomial in the fields so, as the sum converges, $U_t^{(n)}$ is even and in $(\mathscr{A}_n^e)''$ for every $t \in \mathbb{R}$: $\tau^{(n)}(\mathbb{R})$ is an automorphism group of $\mathscr{A}_n^e$. Although misleading in a literal sense, the term 'biquantization' is convenient here: $U_t^{(n)}$ is the biquantization of $u_n^{(1)}(t)$, etc. A similar consideration leads to the semi-group

$$\sigma_\beta^{(n)} = \exp(-\beta \bar{H}_n) \quad (\beta \in \mathbb{R}^+) \tag{2.56}$$

as the biquantization of $\sigma_n^{(1)}(\beta)$. The chain of reasoning leading from eqn (2.53) to eqn (2.55) is typical of Fock-Cook space computation. Details may be found in Cook (1) and Emch [1].

2.7.2. Global time translations

For the whole space $\mathbb{R}^3$ the corresponding situation is as follows. The one-particle Hamiltonian $\bar{h}$ is the unique self-adjoint extension associated with the Laplacian on $\mathbb{R}^3$. Its domain is defined to be

$$\mathbb{D}(\bar{h}) = \left\{ f \in \mathbb{L}^2(\mathbb{R}^3) \ : \ \int_{\mathbb{R}^3} |\widetilde{f}(\xi)|^2 \, \epsilon(\xi) \, \mathrm{d}\xi < \infty \right\}, \tag{2.57a}$$

where $\epsilon(\xi) = \frac{\hbar^2 \ |\xi|^{\ 2}}{2m} - \mu$ is the energy function and $\widetilde{f}$ is the Fourier transform on $\mathbf{L}^2(\mathbf{R}^3)$ (this integral is meant in the mean sense):

$$\widetilde{f}(\xi) = (2\pi)^{-3/2} \int_{\mathbb{R}^3} f(\eta) \exp(-\mathrm{i}\xi \cdot \eta) \, \mathrm{d}\eta. \tag{2.58}$$

For all functions in its domain the formula defining $\bar{h}$ is

$$<f, hg> = \int_{\mathbb{R}^3} [\widetilde{f}(\xi)]^* \, \widetilde{g}(\xi) \, \epsilon(\xi) \, \mathrm{d}\xi, \tag{2.57b}$$

where the star indicates complex conjugation. This formula is sufficient to define the operator $\bar{h}$ because of standard theorems on bilinear forms on Hilbert space; in finite-dimensional cases, they correspond to the equivalence of a linear transformation on a vector space with its matrix element with respect to some basis. The reader is referred to Kato [1] for details.

With this definition of $\bar{h}$ we can write down formulae corresponding to those for $u_n^{(1)}$, $U^{(n)}$, and $\tau^{(n)}$; we shall not consider the analogues of $\sigma_n^{(1)}$ and $\sigma_\beta^{(n)}$. Then we have

$$u_t^{(1)} = \exp(\mathrm{i}t\bar{h}) \quad (t \in \mathbb{R}) \tag{2.59a}$$

for the one particle unitary evolution group on $\mathbb{L}^2(\mathbb{R}^3)$;

$$\tau : \mathbb{R} \to \operatorname{Aut}(\mathscr{A}^\#)$$

$$\tau_t[a^\#(f)] = a^\#[u_t^{(1)}(f)] \quad (\forall f \in \mathbb{L}^2(\mathbb{R}^3)) \tag{2.60a}$$

for the automorphism group on the quasilocal algebras. The unitary evolution group on Fock-Cook space which implements $\tau(\mathbb{R})$ appears through the formula

$$a^\#[u_t^{(1)}(f)] = U_t \, a^\#(f) \, U_{-t} \quad (\forall f \in \mathbb{L}^2(\mathbb{R}^3)) \tag{2.60b}$$

and is given by

$$U_t = \exp(\mathrm{i}t\bar{H}). \tag{2.59b}$$

The Fock-Cook space Hamiltonian, is explicitly,

$$\bar{H} = \sum_{n\in\mathbf{N}}^{(s)} a^*(h^{\frac{1}{2}}\varphi_n)\,a(h^{\frac{1}{2}}\bar{\varphi}_n). \tag{2.61}$$

In this last equation, $\{\varphi_n : n \in \mathbf{N}\}$ is any orthonormal basis of $\mathbf{L}^2(\mathbf{R}^3)$ in the domain of $h^{\frac{1}{2}}$ and the generalized strong sum is meant. That $\tau(\mathbf{R})$ is an automorphism of $\mathscr{A}^{\#}$ follows from eqn (2.60). For this equation shows that $\tau(\mathbf{R})$ maps fields to fields, and hence polynomials in the fields to polynomials. As the polynomials are norm-dense in $\mathscr{A}$ and $\|\tau_t[a^{\#}(f)]\| = \|u_t^{(1)}(f)\| = \|f\|$, τ_t extends continuously to all of $\mathscr{A}^{\#}$.

2.7.3. Convergence of the local time translations

In this subsection it is our intention to quote some standard theorems of analysis and show that the automorphisms $\{\tau^{(n)}(\mathbf{R}) : n \in \mathbf{N}\}$ converge in the sense that

$$\lim_{n\to\infty} \|\tau_t^{(n)}(A) - \tau_t^{(A)}\| = 0 \qquad (A \in \mathscr{A}^{\#}, t \in \mathbf{R}). \tag{2.62a}$$

We choose to write

$$(\text{uniform } \mathscr{A}^{\#}) - \lim_{n\to\infty} \tau_t^{(n)} = \tau_t \quad (\forall t \in \mathbf{R}) \tag{2.62b}$$

for this mode of convergence.

Now the proof. For each $f \in \mathscr{D}(\mathbf{R}^3)$ (the space of infinitely differentiable functions from $\mathbf{R}^3$ to $\mathbf{C}$ of compact support, and equipped with its usual LF-topology† (Choquet [1]; Gel'fand and Shilov [1,2]; Robertson and Robertson [1]; Trèves [1])) there is some integer N such that support of $f \subset V_n$ for all $n > N$. And as $\mathscr{D}(\mathbf{R}^3) \cap \mathbf{L}^2(V_n)$ is dense in $\mathbf{L}^2(V_n)$ (any text, e.g. Trèves [1]), $u_p^{(1)}(t)f$ is well defined for p large enough and $f \in \mathscr{D}(\mathbf{R}^3)$.

Following Gel'fand (Gel'fand and Shilov [1,2]), we shall write $\mathscr{Z}(\mathbf{R}^3)$ for the Fourier image of $\mathscr{D}(\mathbf{R}^3)$; it is the space of entire analytic functions such that $|g(\mathrm{x}+\mathrm{iy})| \leqslant K\exp(-a|\mathrm{y}|)$ $(a > 0)$. Upon noting that $\mathscr{Z}(\mathbf{R}^3) \subset \mathscr{S}(\mathbf{R}^3)$ and that the function $\exp[it\,\epsilon(\xi)]$ is not a multiplier for $\mathscr{Z}(\mathbf{R}^3)$ but is one for $\mathscr{S}(\mathbf{R}^3)$ (Gel'fand and Shilov [1], p.159), it follows that for $t \neq 0$ and omiting the zero function,

$$u_t^{(1)} : \mathscr{D}(\mathbf{R}^3) \to \mathscr{S}(\mathbf{R}^3) \setminus \mathscr{D}(\mathbf{R}^3). \tag{2.63}$$

The difficulty is with the support of the functions; this result is due to the wave-packet dispersion associated with this free Schrödinger equation. This important observation, eqn (2.63), leads to a difference between this model and the Bose model, where the fields are not bounded. For if $a^{\#}(f)$ is unbounded, it cannot be continuously extended from $\mathscr{D}(\mathbf{R}^3)$ to $\mathscr{S}(\mathbf{R}^3)$ with its extension in the quasilocal

† $\mathscr{D}(\mathbf{R}^3)$ is the countable strict inductive limit of the Frechet space $\mathscr{D}(V_n)$ of infinitely differentiable functions with support within V_n. Each $\mathscr{D}(V_n)$ is a metrizable Montel space under the topology defined by the semi-norms $P_m(f) = \sup_{\xi\in V_n} |D^m f(\xi)|$.

The topology on $\mathscr{D}(\mathbf{R}^3)$ is the finest convex topology under which the injection mappings $\mathscr{D}(V_n) \to \mathscr{D}(\mathbf{R}^3)$ are continuous.

algebra. For the Fermi case this does not occur and f may even be in $\mathbb{L}^2(\mathbb{R}^3)$; this is why $\tau(\mathbb{R}) \in \mathrm{Aut}(\mathscr{A})$. But in the boson case, there will not be a $\tau(\mathbb{R})$ automorphism group of the quasilocal algebra, and more care in considering time evolution will be necessary there.

Upon comparing the definitions of $u_n^{(1)}(t)$ and $u_t^{(1)}$, it is clear that

$$(\text{str.}\, \mathbb{L}^2(\mathbb{R}^3)) - \lim_{n\to\infty} u_n^{(1)}(t)f = u_t^{(1)}(f) \qquad (\forall f \in \mathscr{D}(\mathbb{R}^3)). \tag{2.64a}$$

As $\mathscr{D}(\mathbb{R}^3)$ is dense in $\mathbb{L}^2(\mathbb{R}^3)$, this extends to $\mathbb{L}^2(\mathbb{R}^3)$ by continuity. Thus

$$(\text{str.}\, \mathbb{L}^2(\mathbb{R}^3)) - \lim_{n\to\infty} u_n^{(1)}(t)f = u_t^{(1)}(f) \quad (\forall f \in \mathbb{L}^2(\mathbb{R}^3)), \tag{2.64b}$$

where we have identified $\mathbb{L}^2(V_n)$ with a subspace of $\mathbb{L}^2(\mathbb{R}^3)$.

This result combines with the definitions (2.52), (2.53), (2.59), and (2.60) to give a convergence formula for the quantum fields (recall that $j_n^{\#}$ injects $\mathscr{A}_n^{\#}$ into $\mathscr{A}^{\#}$):

$$\| j_n^{\#} \circ \tau_t^{(n)} [a_n^{\#}(f)] - \tau_t [a^{\#}(f)] \| = \| U_n^{(1)}(t)f - U_t^{(1)}(f) \| \to 0. \tag{2.65}$$

But then—and in view of the fact that $\tau(\mathbb{R})$ is an automorphism group of $\mathscr{A}^{\#}$—the uniform convergence result eqn (2.62) quoted above must hold; for the fields form a norm generating set for the algebras. This proves our assertion.

2.8. Computation of the local Gibbs states

Combining eqns (1.35) and (1.36) defining $E_\beta^{(n)}$ and $\phi_\beta^{(n)}$ with the particular semi-group $\sigma_\beta^{(n)}$ for this model (eqn (2.56)) we construct the local Gibbs state $\phi_\beta^{(n)}$. Recall that when we computed the Fock-Cook state $\omega_V^{(0)}$ it sufficed to consider the monomial argument $\mathscr{X}_{pq}$ (eqn (2.31)). What we wish to do is switch over to the spin formalism of § 2.5; in particular, eqns (2.45) and (2.47) show that to do so it is necessary to take elements of the $\mathbb{L}^2(V_n)$ orthonormal basis of eqn (2.44):

$$\{e_\kappa(\xi) = (nL)^{-3/2} \exp(\mathrm{i}\kappa \cdot \xi) : \kappa \in \widetilde{V}_n\}$$

as the functions $f_1, \ldots, g_q$ in $\mathscr{X}_{pq}$. This will lead to considering the argument

$$\mathscr{Y}_{pq} = a_{j_1}^{(n)} \ldots a_{j_p}^{(n)} a_{k_1}^{*(n)} \ldots a_{k_q}^{*(n)}, \tag{2.66}$$

where $j_l, \ldots, k_q \in \mathbb{N}$. Note that to get from $\mathscr{Y}_{pq}$ to more general arguments one can use, e.g.

$$a_n^*(f) = (nL)^{-3/2} \sum_{\kappa \in \widetilde{V}_n} \widetilde{f}_\kappa a_p^{*(n)}, \tag{2.67}$$

where $\eta_n(\kappa) = p$.

Turning now to the computation with $\mathscr{Y}_{pq}$ we introduce the following notation. Two 2×2 projections which we need are

$$Q = \begin{bmatrix} 0 & 0 \\ 0 & 1 \end{bmatrix} \quad \text{and} \quad \bar{Q} = \begin{bmatrix} 1 & 0 \\ 0 & 0 \end{bmatrix}, \tag{2.68}$$

Let us also define

$$F_p^{(n)}(\beta) = \exp\{-\beta\, \epsilon_n[\eta_n^{-1}(p)]\, Q\}, \tag{2.69a}$$

where $\eta_n^{-1}(p) = \kappa \in \widetilde{V}_n$ and $p \in \mathbb{N}$. For then the spin operator image of the semi-group $\sigma_\beta^{(n)}$ is (cf. eqn (2.47))

$$\pi_n[\sigma_\beta^{(n)}] = \bigotimes_{p\in\mathbb{N}} F_p^{(n)}(\beta).$$

Upon expanding the exponential in $F_p^{(n)}(\beta)$ we find that

$$F_p^{(n)}(\beta) = \{\exp[-\beta\, \epsilon_n(\kappa)]\}\, Q + \bar{Q}. \tag{2.69b}$$

There are four sorts of trace calculations to do in order to compute $E_\beta^{(n)}(\mathscr{Y}_{pq})$, namely,

$$J = \mathrm{tr}(\mathrm{e}^{-\lambda Q}) = 1 + \mathrm{e}^{-\lambda}; \tag{2.69c}$$

$$K = \mathrm{tr}(\mathrm{e}^{-\lambda Q} Q) = \mathrm{e}^{-\lambda}; \tag{2.69d}$$

$$L^{(\pm)} = \mathrm{tr}(\mathrm{e}^{-\lambda Q}\sigma^{(\pm)}) = 0; \tag{2.69e}$$

$$L^{(3)} = \mathrm{tr}(\mathrm{e}^{-\lambda Q}\sigma^{(3)}) = 1-\mathrm{e}^{-\lambda}. \tag{2.69f}$$

Now $E_\beta^{(n)}(\mathscr{Y}_{pq})$ will be a product of such quantities, but with mode subscripts attached, e.g. $K_i = \exp[-\beta\, \epsilon_n(\eta_n^{-1}(i)]$. A check will reveal that every $L_i^{(3)}$ which appears will be multiplied by at least one $L_j^{(\pm)} = 0$, and hence the $L_i^{(\alpha)}$ can safely be ignored. In fact a little more is easily seen: $E_\beta^{(n)}(\mathscr{Y}_{pq})$ will vanish unless $(j_1, \ldots, j_p)$ is some cyclic permutation of $(i_1, \ldots, i_q)$. The easiest case is $p=q=0$, by which we mean $\mathscr{Y}_{00} = \mathbb{1}$; this is the grand partition function and consists entirely of the Js:

$$E_\beta^{(n)}(\mathbb{1}) = \prod_{\kappa\in\widetilde{V}_n} \{1 + \exp[-\beta\, \epsilon_n(\kappa)]\}, \tag{2.70}$$

which is a more-or-less familiar result. What we must first show is that this product actually exists (converges). But using $\epsilon_n(\kappa) = \frac{\hbar^2}{2m}|\kappa|^2 - \mu_n$, it is a simple estimate that

$$\sum_{\kappa\in\widetilde{V}_n} \log\{1 + \exp[-\beta\, \epsilon_n(\kappa)]\} < \infty, \tag{2.71}$$

and this ensures that $E_\beta^{(n)}(1) < \infty$. In turn this finally proves that $\sigma_\beta^{(n)}$ is trace-class.

For $p = q > 0$ the non-zero contributions to $E_\beta^{(n)}(\mathscr{Y}_{pq})$ come when $I_p = (i_1, \ldots, i_p) = g(j_1, \ldots, j_p)$, where g is some element of the permutation group on p letters. If $|g|$ denotes the parity of the permutation,

$$E_\beta^{(n)}(\mathscr{Y}_{pq}) = |g|\,(\prod_{j\in I_p} K_j)(\prod_{k\in\mathbb{N}\setminus I_p} J_k),$$

whence

$$\phi_\beta^{(n)}(\mathscr{Y}_{pp}) = |g| \prod_{j\in I_p} (1 + \exp\{\beta\, \epsilon_n[\eta_n^{-1}(j)]\}). \tag{2.72}$$

Going back to $\mathscr{X}_{pq}$ yields

$$\phi_\beta^{(n)}(\mathscr{X}_{pq}) = \delta_{pq} \det|\phi_\beta^{(n)}[a_n(\bar{f}_i)a_n^*(g_k)]|, \tag{2.73}$$

with two-point function

$$\phi_\beta^{(n)}[a_n(\tilde{f})a_n^*(g)] = (nL)^{-3} \sum_{\kappa\in\tilde{V}_n} \tilde{f}(\kappa)^*\tilde{g}(\kappa)\{1+\exp[\beta\epsilon_n(\kappa)]\}^{-1} \tag{2.74}$$

$$= \langle f, \rho_n^{(-)}(\beta)\, g\rangle. \tag{2.75}$$

We have introduced $\rho_n^{(-)}(\beta)$, the Fermi-Dirac density operator on $\mathbb{L}^2(\mathbb{R}^3)$ defined by

$$\rho_n^{(-)}(\beta) = \sum_{\kappa\in\tilde{V}_n} \{1+\exp[\beta\epsilon_n(\kappa)]\}^{-1} e_\kappa \otimes e_\kappa^* \tag{2.76}$$

$$= [1+\sigma_n^{(1)}(\beta)]^{-1}. \tag{2.77}$$

The local Gibbs state is quasi-free, being a determinant of its two-point functions; we shall not distinguish between $\phi_\beta^{(n)} \in \mathfrak{S}(\mathscr{A}_n)$ and its restriction to $\mathscr{A}_n^{\mathrm{e}}$ just as for the Fock-Cook state, since it vanishes on the set $\mathscr{A}\backslash\mathscr{A}_n^{\mathrm{e}}$.
It is clearly gauge invariant, but it is not translationally invariant. However, whilst we must logically distinguish $\phi_\beta^{(V_n)}$ from $\phi_\beta^{(V_n+\xi)}$, they are numerically equal, by which we mean that the simultaneous replacement $\tilde{f}(\kappa) \to \exp(\mathrm{i}\kappa\cdot\xi)\tilde{f}(\kappa)$ and $\tilde{g}(\kappa) \to \exp(\mathrm{i}\kappa\cdot\xi)\tilde{g}(\kappa)$ cancel each other out.

Sirugue and Testard (1) have shown that for all functions in the set

$$\mathscr{V}^{(n)} = \left\{\int_{\mathbb{R}} \mathrm{d}t\, f(t)\, u_n^{(1)}(t)\, g(t) : f\in \mathscr{Z}(\mathbb{R}), g\in \mathbb{L}^2(V_n)\right\}, \tag{2.78a}$$

the expressions

$$\tau_{t+\mathrm{i}\beta}^{(n)}[a_n^{\#}(f)] = a_n^{\#}\{u_n^{(1)}(t+\mathrm{i}\beta)[f]\} \quad (f\in\mathscr{V}^{(n)}) \tag{2.78b}$$

make good sense; and that $\mathscr{V}^{(n)}$ is dense in $\mathbb{L}^2(V_n)$. This being so, we can prove that $\phi_\beta^{(n)}$ is $(\beta, \tau^{(n)})$-KMS as follows: for $f\in\mathbb{L}^2(V_n)$, $g\in\mathscr{V}^{(n)}$, one has

$$\phi_\beta^{(n)}\left(a_n[u_n^{(1)}(t)g]\, a_n^*(f)\right)$$

$$= \langle f, u_n^{(1)}(t)g\rangle - \phi_\beta^{(n)}\left(a_n^*(f)\, a_n[u_n^{(1)}(t)g]\right)$$

$$= \sum_{\kappa\in\tilde{V}_n} \tilde{f}(\kappa)^*\tilde{g}(\kappa)\exp[-\mathrm{i}t\,\epsilon_n(\kappa)]\left(1-\{1+\exp[\beta\epsilon_n(\kappa)]\}^{-1}\right)$$

$$= \phi_\beta^{(n)}\left(a_n^*(f)\, a_n\{u_n^{(1)}(t+\mathrm{i}\beta)[g]\}\right).$$

It can be seen that the derivation of eqns (2.73)-(2.75) pertaining to $\phi_\beta^{(n)}$ is exact: any approximation in the theory is in the choice of the Hamiltonian and in the use of $\phi_\beta^{(n)}$. Eqn (2.76) shows the famous Fermi-Dirac factor $[1+\exp(\beta\epsilon_n)]^{-1}$. The operator $\rho_n^{(-)}(\beta)$ is a particle density operator, as (2.74)-(2.75) show that the mean density in the state $\phi_\beta^{(n)}$ is $\bar{\rho} = \mathrm{tr}[\rho_n^{(-)}(\beta)]$, since the number operator is a sum over orthonormal modes (eqn (2.19)). That is, and this will be amplified in the next section,

$$\bar{\rho} = (nL)^{-3} \sum_{\kappa\in\tilde{V}_n} \{1+\exp[\beta\epsilon_n(\kappa)]\}^{-1}.$$

There is *no* divergence associated with the fact that in the thermodynamic limit '$\epsilon_n(\kappa) \to \epsilon(k)$' with $\epsilon(k)$ taking the value zero at $k = 0$ for zero chemical potential. But this is not the case for bosons, where the minus sign occurs; the consequent divergence associated with the particle density is known as the *Bose-Einstein condensation*, and we shall show in Chapter 3 that it is indicative of a phase transition. Thus the different signs (±) in the density operator leads to the presence or absence of a phase transition. One may say that the exclusion principle prevents free electrons from joining together in a collective state to cause a phase transition; for the exclusion principle implies the Fock-Cook space antisymmetry which in turn leads to *anti*commutation relations; and the *anti*commutation relations are at the source of the + sign in $\rho_n^{(-)}(\beta)$ as the following well-known heuristic argument shows. Using the KMS condition for the pth mode:

$$\phi_\beta^{(n)}[\tau_t(a_\kappa^*)a_\kappa] = \phi_\beta^{(n)}[a_\kappa \tau_{t+i\beta}(a_\kappa^*)];$$

this may be rewritten as

$$\exp[it\epsilon_n(\kappa)] <N_\kappa>_\beta = \exp[i(t+i\beta)\epsilon_n(\kappa)] <a_\kappa a_\kappa^*>_\beta,$$

so that

$$<N_\kappa>_\beta = \exp[-\beta\epsilon_n(\kappa)] <1-N_\kappa>_\beta;$$

or finally

$$<N_\kappa>_\beta = \{1+\exp[+\beta\epsilon_n(\kappa)]\}^{-1}.$$

We see from this how each Fermi-Dirac oscillator, so-called, contributes to the density; here $<N_\kappa>_\beta \equiv \phi_\beta^{(n)}[a_\kappa^* a_\kappa]$ for brevity. Even at this point, disregarding all questions of rigour, the corresponding boson calculation is possible; the difference is that $<1-N_\kappa>_\beta$ would be replaced by $<1+N_\kappa>_\beta$ from the CCR (see eqns (3.9a) and (3.9b)); and consequently $<N_\kappa>_\beta$ would equal $\{1-\exp[\beta\epsilon_n(\kappa)]\}^{-1}$.

2.9. The global Gibbs state

Having just computed the local Gibbs states $\phi_\beta^{(n)}$ (eqns (2.73)-(2.77)) we now wish to find the limit $\phi_\beta^{(n)} \to \phi_{\beta\bar{\rho}}$ subject to the constraint $\phi_\beta^{(n)}[N_n/(nL)^3] = \bar{\rho}$ appropriate to our grand canonical formalism. Two initial remarks: our limiting state corresponds to uniform prescribed density $\bar{\rho}$; our limiting state corresponds to the particular choices of a net of cubes $\mathcal{M}$ and cyclic boundary conditions. These restrictions can be relaxed if one is willing to do extra work: see Lewis and Pulé (1). Historically, the method employed here is due to Kac by way of Lewis and Pulé and Cannon (1). The first result on such a limit state is due to Araki and Wyss (1), as mentioned before. If it were only the Fermi gas that we wished to consider, our method would be too involved. But we shall consider the Bose gas as well, and have chosen to use the same approach for both. It is instructive to see the prescribed density constraint made explicit, even in this Fermi gas model where it is not as important as for the Bose gas.

By

$$z_n = \exp(\beta\mu_n) \tag{2.79}$$

we shall mean the fugacity—sometimes known as the (chemical) activity—for the local region V_n. The eigenvalues of the one-particle non-reduced Hamiltonian, i.e. without the subtraction of the chemical potential, are $\omega_n(\kappa) = \frac{\hbar^2\,|\kappa|^2}{2m}$ $(\kappa \in \widetilde{V}_n)$. In eqn (2.82), the activity z_n will be related to the prescribed density $\bar{\rho}$, so it will be advantageous to separate out the fugacity contributions.

For real $\xi \in \mathbb{R}$ let us define the Fourier transform of the Kac density:

$$\widetilde{K}_n^{(-)}(\xi, \bar{\rho}) = \phi_{\beta\bar{\rho}}^{(n)}\,[\exp(-\mathrm{i}\xi N_n/\,|\,V_n|)], \tag{2.80a}$$

i.e. the Fourier inverse of $\widetilde{K}_n^{(-)}$, written $K_n^{(-)}$, will be called the Kac density (Lewis (1); Lewis and Pulé (1); Cannon (1)). The reader will notice the density symbol appended to the Gibbs state. This density function can be calculated by means of the spin formalism. As the calculation is quite similar to the one for the grand partition function we omit it; the result is that

$$\widetilde{K}_n^{(-)}(\xi,\bar{\rho})$$

$$= \prod_{\kappa\in\widetilde{V}_n} \{1+z_n\exp[-\beta\,\omega_n(\kappa)]\}\;\{1+z_n\exp[-\beta\,\omega_n(\kappa)+(\mathrm{i}\,\xi/\,|\,V_n|)]\}^{-1}. \tag{2.80b}$$

As $\phi_{\beta\bar{\rho}}^{(n)}$ is quasi-free it suffices to consider the correlation function

$$\phi_{\beta\bar{\rho}}^{(n)}\,[\exp(\mathrm{i}\,\xi N_n/\,|\,V_n|)\,a_n^*(f)\,a_n(\bar{g})]$$

$$= (nL)^{-3}\widetilde{K}_n^{(-)}(\xi,\bar{\rho}) \sum_{\kappa\in V_n} z_n \widetilde{f}(\kappa)\widetilde{g}(\kappa)^* \times$$

$$\times\{z_n + \exp[\beta\,\omega_n(\kappa) + (\mathrm{i}\xi/\,|\,V_n|)]\}^{-1}. \tag{2.81}$$

The relation between density and activity—which we shall refer to as the intensive equation of state—is now evident;

$$\bar{\rho} = \phi_{\beta\bar{\rho}}^{(n)}\,[N_n/\,|\,V_n|]$$

$$= \mathrm{i}\,\frac{\partial}{\partial\xi}\,\widetilde{K}_n^{(-)}(\xi,\bar{\rho})\Big|_{\xi=0},$$

or

$$\bar{\rho} = (nL)^{-3} \sum_{\kappa\in\widetilde{V}_n} z_n\{z_n + \exp[\beta\,\omega_n(\kappa)]\}^{-1}. \tag{2.82}$$

The crux of the proof of the convergence of the relevant quantities in the thermodynamic limit is contained in the following lemma.

LEMMA 2.1

Let $G_\beta^{(-)} : \mathbb{R}^+ \to \mathbb{R}$ *be the function defined by*

$$G_\beta^{(-)}(x) = (2\pi)^{-3} \int_{\mathbb{R}^3} x\,\{x + \exp[\beta\,\omega\,(\mathrm{k})]\}^{-1}\,\mathrm{dk}$$

$$= (2\pi^2)^{-1} \int_0^\infty xk^2 \{x + \exp[\beta\omega(\mathrm{k})]\}^{-1} \, \mathrm{dk}, \tag{2.83}$$

where $\omega(\mathbf{k}) = \hbar^2 |\mathbf{k}|^2/2m$. *Then*

$$\lim_{n\to\infty} (nL)^{-3} \sum_{\kappa\in\widetilde{V}_n} x\{x + \exp[\beta\omega_n(\kappa)]\}^{-1} = G_\beta^{(-)}(x) \tag{2.84}$$

uniformly in $x \in \mathbb{R}^+$.

Proof. Cannon (1) shows that the terms in $\kappa \in \widetilde{V}_n$ for which any component κ_j $(j = 1,2,3)$ vanishes will vanish uniformly in the limit. The other terms are shown to be step functions which approximate the integrand of $G_\beta^{(-)}$ and vanish uniformly outside some neighbourhood of the origin in the $\mathbb{R}^+ \times V_n$ plane, i.e. $(x,\kappa) = (0,0)$. This completes the proof.

We shall suppose that the fugacity z_n approaches a limit: $z_n \to z_\infty$ with $z_\infty \in \mathbb{R}^+$ as n approaches infinity. The left-hand side of eqn (2.84) is just the expression eqn (2.82) which we have found for the mean density; the choice of $x = z_\infty$ is possible since the convergence is uniform in the fugacity. Thus we have as a corollary to Lemma 2.1 the equation

$$\bar{\rho} = G_\beta^{(-)}(z_\infty), \tag{2.85}$$

which is the thermodynamical intensive equation of state, as it relates the temperature, density, and fugacity. Moreover, as $G_\beta^{(-)}$ is uniquely invertible to find $z_\infty(\beta,\bar{\rho})$, the fugacity is uniquely determined by the density for all temperatures. Thus, either $(\beta,\bar{\rho})$ or (β,z_∞) can be used as independent thermodynamic variables in their entire ranges. In the Bose gas model matters are not so simple as we shall see. Let us separate the lowest $\kappa = 0$ term for the density explicitly:

$$\bar{\rho} = (nL)^{-3} z_n(1+z_n)^{-1} + (nL)^{-3} \sum_{\kappa\neq 0,\, \kappa\in\widetilde{V}_n} z_n\{z_n + \exp[\beta\omega_n(\kappa)]\}^{-1}. \tag{2.86}$$

We see that the first term converges to zero and the modes $\kappa \in V_n$, $\kappa \neq 0$ give the total density. In the Bose case, the corresponding first term is $(nL)^{-3} z_n/(1-z_n)$ which looks as though it might diverge for $z_n \to 1$. The dependence of z_n on n is determined by the requirement that the density be fixed and this first term will be seen to converge to the superfluid density equation (3.45) (p. 60).

In order to examine the limit $\phi_{\beta\bar{\rho}}^{(n)} \to \phi_{\beta\bar{\rho}}$ we must find the limit of $\widetilde{K}_n^{(-)}$. The pertinent lemma is due to Kac:

LEMMA 2.2

$$\lim_{n\to\infty} \widetilde{K}_n^{(-)}(\xi,\bar{\rho}) = \exp[\mathrm{i}\xi G_\beta^{(-)}(z_\infty)]$$
$$= \exp(\mathrm{i}\xi\bar{\rho}) \tag{2.87}$$

for each $\xi \in \mathbb{R}^+$ *uniformly on bounded sets.*

Proof. We start from the expression

$$\widetilde{K}_n^{(-)}(\xi,\bar{\rho}) = \exp\Big[-\sum_{\kappa\in\widetilde{V}_n} \log\Big(\{1+z_n\exp[-\beta\omega_n(\kappa)-(\mathrm{i}\xi/|V_n|)]\}\times$$
$$\times\{1+z_n\exp[-\beta\omega_n(\kappa)]\}^{-1}\Big)\Big].$$

In the numerator of the argument of the logarithm we expand $\exp[-\mathrm{i}\xi/|V_n|] = 1-\mathrm{i}\xi/|V_n|$ and higher orders (h.o.) in n^{-1}. Call $1+z_n\exp(-\beta\omega_n)$ the symbol a_n, for brevity; then the argument of the logarithm becomes $\simeq\{a_n+[\mathrm{i}\xi/|V_n|] \times z_n\exp(-\beta\omega_n)\}/a_n$, or upon dividing through by a_n and $\exp(-\beta\omega_n)$

$$1-\mathrm{i}\xi(nL)^{-3}z_n[z_n+\exp(\beta\omega_n)]^{-1}+\text{h.o.}(n^{-1}).$$

The logarithm of this is expanded in accordance with $\log(1+\gamma)\simeq\gamma-\frac{1}{2}\gamma^2 + \text{h.o.}(\gamma)$ to give

$$\widetilde{K}_n^{(-)}(\xi,\bar{\rho}) = \exp\Big(\sum_{\kappa\in\widetilde{V}_n}\mathrm{i}\xi(nL)^{-3}z_n\{z_n+\exp[\beta\omega_n(\kappa)]\}^{-1}+\text{h.o.}(n^{-1})\Big).$$

From the continuity of the exponential and the uniform convergence of the limit in eqn (2.84) we may use (2.84) and this above expression to immediately infer eqn (2.87). This completes the proof.

Just as mentioned above for $\bar{\rho}$, the $\kappa=0$ term converges simply for this model:

$$\widetilde{K}_n^{(-)}(\xi,\bar{\rho})\to\frac{1+z_\infty}{1+z_\infty}\exp[\mathrm{i}\xi G_\beta^{(-)}(z_\infty)]$$
$$=\exp[\mathrm{i}\xi G_\beta^{(-)}(z_\infty)],$$

but must be analysed more carefully for the Bose gas.

Putting these facts together leads to the main theorem.

THEOREM 2.1

For any $\beta,\bar{\rho}$ and every $f,g\in\mathscr{D}(\mathbb{R}^3)$,

$$\phi_{\beta\bar{\rho}}^{(n)}\{\exp(-\mathrm{i}\xi N_n/|V_n|)\,a_n^*(f)\,a_n(\bar{g})\}\to\exp(-\mathrm{i}\xi\bar{\rho})\,\phi_{\beta\bar{\rho}}[a^*(f)\,a(\bar{g})] \tag{2.88}$$

uniformly in bounded sets in $\xi\in\mathbb{R}^+$, where (cf. eqn (2.75))

$$\phi_{\beta\bar{\rho}}[a^*(f)\,a(\bar{g})] = \langle f,\rho_\beta^{(-)}g\rangle, \tag{2.89}$$

where $\rho_\beta^{(-)}$ is defined by

$$(\rho_\beta^{(-)}f)^\sim(\mathbf{k}) = \Big\{1+(z_\infty)^{-1}\exp\Big(\frac{\beta\hbar^2}{2m}|\mathbf{k}|^2\Big)\Big\}^{-1}\widetilde{f}(\mathbf{k})$$
$$\equiv\rho_\beta^{(-)}(\mathbf{k},z_\infty)\widetilde{f}(\mathbf{k}). \tag{2.90}$$

Proof. From eqn (2.81) the limit

$$\phi_{\beta\bar{\rho}}^{(n)}\big[\exp(-\mathrm{i}\xi N_n/|V_n|)\,a_n^*(f)\,a_n(\bar{g})\big]$$

$$\rightarrow \lim_{n\to\infty} \widetilde{K}_n^{(-)}(\xi,\bar{\rho}) \lim_{n\to\infty} (nL)^{-3} \sum_{\kappa\in\widetilde{V}_n} z_n \widetilde{f}(\kappa)\widetilde{g}^*(\kappa) \times$$

$$\times \{z_n + \exp[\beta\omega_n(\kappa) + \mathrm{i}(\xi/|V_n|)\}^{-1}$$

$$= \exp(\mathrm{i}\xi\bar{\rho}) \lim_{n\to\infty} (nL)^{-3} \sum_{\kappa\in\widetilde{V}_n} z_n \widetilde{f}(\kappa)\widetilde{g}(\kappa)^* \{z_n + \exp[\beta\omega_n(\kappa) + \mathrm{i}(\xi/|V_n|)]\}^{-1}.$$

is a matter of the definition of the Lebesgue integral. For the quantity in braces converges to the integral $\langle f, \rho_\beta^{(-)} g\rangle$, using (2.90). The restriction of f and g to $\mathscr{D}(\mathbf{R}^3)$ ensures the vanishing of the summands outside some compact interval in $\mathbf{R}^3$, and so outside some neighbourhood of the $\mathbf{R}^+ \times \mathbf{R}^3$ origin $(\xi,\kappa) = (0,0)$. The uniformity follows from this. This completes the proof.

As $\phi_{\beta\bar{\rho}}$ corresponds to the density $\bar{\rho}$ it is locally normal. From the explicit form given in (2.89) and (2.90), it is evident that $\phi_{\beta\bar{\rho}}$ is gauge invariant and space-translationally invariant. The GNS representation associated with $\phi_{\beta\bar{\rho}}$ will show that the representations are factorial and later we see that they are the unique KMS states. This enables us to conclude that there is no phase transition and no symmetry breakdown.

2.10. The thermodynamic representations

By the thermodynamic representations we mean the canonical triple $\phi_{\beta\bar{\rho}} \sim [\mathscr{H}_{\beta\bar{\rho}}, \pi_{\beta\bar{\rho}}, \Omega_{\beta\bar{\rho}}]$. The best course would seem to be just to write the triple down: the interested reader can then verify their validity. We do this because we do not know a derivation of them in the form we give them. Recalling then that $\mathscr{H}$ is the Fock-Cook Hilbert space $\mathscr{H}_{\mathrm{F}}(\mathbf{R}^3)$ constructed over $\mathbf{L}^2(\mathbf{R}^3)$ from antisymmetric tensors, $\mathscr{H}_{\beta\bar{\rho}}$ is given by

$$\mathscr{H}_{\beta\bar{\rho}} = \mathscr{H} \otimes \mathscr{H}, \tag{2.91a}$$

with canonical cyclic vector

$$\Omega_{\beta\bar{\rho}} = \Omega_0 \otimes \Omega_0 \tag{2.91b}$$

constructed out of the Fock-Cook vacuum $\Omega_0 \equiv \Omega_{\mathbf{R}^3} \in \mathscr{H}$.

The representation itself follows from the following formula, where $a^\#$ and a^b are either a or a^* constrained so that $\# \neq b$,

$$\pi_{\beta\bar{\rho}}[a^\#(f)] = a^\#([1-\rho_\beta^{(-)}]^{\frac{1}{2}}f) \otimes \mathbb{1} - \exp[\mathrm{i}\tfrac{\pi}{2}N(\mathbf{R}^3)] \otimes a^b([\rho_\beta^{(-)}]^{\frac{1}{2}}f), \tag{2.91c}$$

where $N(\mathbf{R}^3)$ is the number operator on $\mathscr{H}$. The representation algebra is a proper subalgebra

$$\mathscr{A}_{\beta\bar{\rho}} = \pi_{\beta\bar{\rho}}(\mathscr{A}) \subset \mathscr{A} \otimes \mathscr{A}, \tag{2.91d}$$

of the C^*-tensor product of the quasilocal algebra with itself. The restriction from $\mathscr{A}$ to $\mathscr{A}^{\mathrm{e}}$ is obvious.

Dell'Antonio (1) has shown that $\pi_{\beta\bar{\rho}}$ is equivalent to $\pi_{\beta'\bar{\rho}'}$ if and only if $\beta = \beta'$ and $\bar{\rho} = \bar{\rho}'$ and that each algebra $\mathscr{A}''_{\beta\bar{\rho}}$ is a Type-III factor.

A result akin to this was obtained by Powers (1, 2), whose work is easiest to describe in the spin formalism. The algebra in question is therefore $\mathfrak{B}$; and let $\varphi \in \mathfrak{S}(\mathfrak{B})$ be the product state $\varphi = \bigotimes_{\mathbb{N}} \varphi_n$, where φ_n is defined by

$$\varphi_n[a\mathbb{1} + bJ_n^{(1)} + cJ_n^{(2)} + dJ_n^{(3)}] = p_n(a+d) + (1-p_n)(a-d). \qquad (2.92)$$

If $p_n = 0$ for every n, $\pi_\varphi(\mathfrak{B})''$ is a Type-I_∞ factor. For $p_n = \lambda$, $0 < \lambda < \frac{1}{2}$, $\pi_\varphi(\mathfrak{B})''$ is a Type-III factor. It is customary to denote $\pi_\varphi(\mathfrak{B})'' = M_\lambda$ and call it the λ *Power's factor*. The importance of this is that each M_λ is not *-isomorphic to M_σ for $\lambda \neq \sigma$. It is because each M_λ is a representation of the CAR that we mention this here.

Although we have not directly computed the convergence—or no—of the canonical states, there is reason to believe that they do converge, by analogy with the Bose gas. These canonical states are defined as follows.

Let $P_V^{(n)} : \mathscr{H}_{\mathrm{F}}(V) \to \mathbb{L}_{\mathrm{A}}^2(V^n)$ be the Fock-Cook space projection operator indicated in eqn (2.4). For fixed n and V, let $\rho = n/|V|$. Let $\Xi(\mathrm{V},\beta)$ be the semi-group $\exp[-\beta H(V)]$ formed with the non-reduced Hamiltonian on $\mathscr{H}_{\mathrm{F}}(V)$.† Then the canonical state $\psi_{\beta\rho}^{(V)} \in \mathfrak{S}[\mathscr{A}(V)]$ is given by

$$\psi_{\beta\rho}^{(V)}(A) = \frac{\mathrm{tr}_V[\Xi(V,\beta)\,P_V^{(\rho V)}\,A]}{\mathrm{tr}_V[\Xi(V,\beta)\,P_V^{(\rho V)}]}. \qquad (2.93a)$$

Such states are defined for densities at the points $\rho = n/|V|$ only. For densities ρ'' between $\rho = n/V$ and $\rho' = (n+1)/|V|$ $\psi_{\beta\rho}^{(V)}$ is defined by interpolation,

$$\psi_{\beta\rho''}^{(V)} = \lambda\,\psi_{\beta\rho}^{(V)} + (1-\lambda)\,\psi_{\beta\rho}^{(V)} \quad (0 \leqslant \lambda = \lambda(\rho'') \leqslant 1). \qquad (2.93b)$$

Consider the expression ($V \equiv V_n$)

$$\begin{aligned}\Lambda_j(A) &= \psi_{\beta\, j/|V|}^{(V)}(A)\,\phi_{\beta\bar{\rho}}(P_V^{(j)}) \\ &= \frac{\mathrm{tr}[\Xi(V,\beta)P_V^{(j)}A]\,\mathrm{tr}[\Xi(V,\beta)\,z^{N(V)}P_V^{(j)}]}{[\mathrm{tr}[\Xi(V,\beta)P_V^{(j)}][\mathrm{E}_\beta^{(n)}(\mathbb{1})]}\end{aligned}$$

Proceeding formally, $N(V) = \Sigma\, j P_V^{(j)}$, and for any $B \in \mathscr{A}_n^{\#}$,

$$\begin{aligned}\mathrm{tr}(B\,z^{N(V)}P_V^{(j)}) &= z^j\,\mathrm{tr}\left\{B \prod_{k\neq j}[z^k P_V^{(k)} + (1-P_V^{(k)})]\,P_V^{(j)}\right\} \\ &= z^j\,\mathrm{tr}(BP_V^{(j)}),\end{aligned}$$

whence

$$\begin{aligned}\Lambda_j(A) &= z^j\,\mathrm{tr}[\Xi(V,\beta)P_V^{(j)}A]/E_\beta^{(n)}(\mathbb{1}). \\ &= \mathrm{tr}[\Xi(V,\beta)\,z^{N(V)}P_V^{(j)}A]/E_\beta^{(n)}(\mathbb{1}).\end{aligned}$$

† This is defined as $\bar{H}(V)$ for zero chemical potential (cf. eqns (1.28) and (2.55)).

Upon summing over j to a finite limit, say J, the trace and sum may be interchanged. The limit $J \to \infty$ then converges to give

$$\lim_{J\to\infty} \sum_{j=1}^{J} \Lambda_j(A) = \mathrm{tr}\left\{\Xi(V,\beta)\, z^{N(V)} \left(\sum_{j=1}^{\infty} P_V^{(j)}\right) A\right\} / E_\beta^{(n)}(\mathbb{1})$$
$$= \phi_{\beta\bar{\rho}}^{(n)}(A).$$

This then relates the canonical and grand canonical states:

$$\phi_{\beta\bar{\rho}}^{(n)}(A) = \sum_{j\in\mathbf{N}} \phi_{\beta\bar{\rho}}^{(n)}(P_{V_n}^{(j)})\, \psi_{\beta\, j/|V_n|}^{(V_n)}(A). \tag{2.94}$$

The role of the Kac density becomes clear upon noting that

$$\exp\left[-\mathrm{i}\frac{\xi}{|V_n|} N_n\right] = \sum_{j\in\mathbf{N}} \exp\left[-\mathrm{i}\xi \frac{j}{|V_n|}\right] P_{V_n}^{(j)},$$

for then the definition of $\widetilde{K}^{(-)}$ can be rewritten in the form

$$\widetilde{K}^{(-)}(\xi,\bar{\rho}) = \phi_{\beta\bar{\rho}}^{(n)}[\exp(-\mathrm{i}\xi N_n/|V_n|)]$$
$$= \sum_{j\in\mathbf{N}} \exp(-\mathrm{i}\xi j/|V_n|)\, \phi_{\beta\bar{\rho}}^{(n)}(P_{V_n}^{(j)}),$$

from which we recognize that

$$K^{(-)}\left(\frac{j}{|V_n|},\bar{\rho}\right) = \phi_{\beta\bar{\rho}}^{(n)}(P_{V_n}^{(j)}), \tag{2.95}$$

which gives the explicit form of $K^{(-)}$. Upon combining (2.94) and (2.95) we can write

$$\phi_{\beta\bar{\rho}}^{(n)}(A) = \sum_{j=\mathbf{N}} K^{(-)}\left(\frac{j}{|V_n|}, \bar{\rho}\right) \psi_{\beta\, j/|V_n|}^{(V_n)}(A). \tag{2.96}$$

Let us assume that the canonical states converge; no one has shown this for the ideal Fermi gas but as Cannon (1) has shown it for the ideal Bose gas one can presumably adapt his proof here. The limit will be written $\psi_{\beta\rho}(\mathbb{R}^3) \equiv \psi_{\beta\rho}$, and then (2.96) together with Lemma 2.2 gives

$$\phi_{\beta\bar{\rho}}(A) = \int_{\mathbb{R}^+} \mathrm{d}\rho K^{(-)}(\rho,\bar{\rho})\, \psi_{\beta\rho}(A), \tag{2.97}$$

where $K^{(-)}$ is the Kac density. For this model, $K^{(-)}$ in the limit may be explicitly computed; it is

$$K^{(-)}(\rho,\bar{\rho}) = \int_{\mathbb{R}^+} \exp[-\mathrm{i}\xi\rho]\, \widetilde{K}^{(-)}(\xi,\bar{\rho})\, \mathrm{d}\xi/2\pi$$
$$= \delta(\rho - \bar{\rho}). \tag{2.98}$$

But this means that, modulo convergence,

$$\phi_{\beta\bar{\rho}} = \psi_{\beta\bar{\rho}}. \tag{2.99}$$

It would be worthwhile to compute independently the microcanonical state and show that it is equal to the canonical state. Such a proof does not yet exist and constitutes a definite gap in the analysis of this model.

2.11. The Liouville equation

We have already found that the global dynamics for the algebra is given through the automorphism group $\tau : \mathbb{R} \to \mathrm{Aut}(\mathscr{A})$. We now use $\tau(\mathbb{R})$ to examine the dynamics on $\mathscr{H}_{\beta\bar{\rho}}$. Writing

$$a^{\#}_{\beta\bar{\rho}}(f) = \pi_{\beta\bar{\rho}}[a^{\#}(f)],$$

we define the representation automorphism group

$$\tau_{\beta\bar{\rho}} : \mathbb{R} \to \mathrm{Aut}(\mathscr{A}_{\beta\bar{\rho}})$$

$$\tau_{\beta\bar{\rho}}(t) : a^{\#}_{\beta\bar{\rho}}(f) \mapsto a^{\#}_{\beta\bar{\rho}}[u^{(1)}_t(f)]; \tag{2.100}$$

and

$$\tau_{\beta\bar{\rho}}(t) : Q \mapsto U_{\beta\bar{\rho}}(t)\, Q\, U_{\beta\bar{\rho}}(t) \quad (Q \in \mathscr{A}_{\beta\bar{\rho}}). \tag{2.101}$$

The formal generator of $U_{\beta\bar{\rho}}$ is the Liouville operator

$$L_{\beta\bar{\rho}} = \sum a^*{}_{\beta\bar{\rho}}(e_n)\, a_{\beta\bar{\rho}}(\bar{h}\bar{e}_n) \qquad \text{(formal)}, \tag{2.102a}$$

which acts by derivation, i.e. through commutations. Here $\{e_n\}$ is some orthonormal basis of $\mathbb{L}^2(\mathbb{R}^3)$ and $\bar{h}$ is the one-particle Hamiltonian. The spectrum of $L_{\beta\bar{\rho}}$ is the whole real line:

$$\mathrm{spec}(L_{\beta\bar{\rho}}) = \mathbb{R}. \tag{2.102b}$$

Thus we have a *quantum Liouville equation:*

$$[L_{\beta\bar{\rho}}, Q]_- = \mathrm{i}^{-1}\, \text{str.-}\lim_{t\to 0} \tfrac{\mathrm{d}}{\mathrm{d}t}[\tau_{\beta\bar{\rho}}(t) Q] \quad (Q \in \mathscr{A}_{\beta\bar{\rho}}), \tag{2.103}$$

sometimes abbreviated

$$[L_{\beta\bar{\rho}}, Q]_- = \mathrm{i}^{-1}\dot{Q}.$$

2.12. The KMS condition

Our final task for this Fermi gas is to show that $\phi_{\beta\bar{\rho}}$ is $(\beta, \tau_{\beta\bar{\rho}})$-KMS. It is sufficient to consider two point functions. Define then

$$\mathscr{X} = \int_{\mathbb{R}} F(t - \mathrm{i}\beta)\, \phi_{\beta\bar{\rho}}\{a^*(f)\, a[u^{(1)}_t(\bar{g})]\}\, \mathrm{d}t \tag{2.104a}$$

and

$$\mathscr{Y} = \int_{\mathbb{R}} F(t)\, \phi_{\beta\bar{\rho}}\, a[u^{(1)}_t(\bar{g})]\, a^*(f)\ \mathrm{d}t \tag{2.104b}$$

for arbitrary $F \in \mathscr{Z}(\mathbb{R})$. The explicit form of $\phi_{\beta\bar{\rho}}$ leads to

$$\mathscr{X} = \int_{\mathbb{R}} F(t - \mathrm{i}\beta) < f, \rho^{(-)}_\beta\, u^{(1)}_t(g) > \mathrm{d}t \tag{2.105a}$$

and

$$\mathscr{Y} = \int_{\mathbb{R}} F(t) < f, (1 - \rho^{(-)}_\beta)\, u^{(1)}_t(g) > \mathrm{d}t; \tag{2.105b}$$

the CAR have been used here.

Granting a proof of the requisite continuity so as to invoke Fubini's theorem, we write out the inner product and interchange the $\mathbb{R}$ and $\mathbb{R}^3$ integrations in $\mathscr{X}$:

$$\mathscr{X} = \int_{\mathbb{R}^3} f(\mathbf{k})\, g(\mathbf{k})^* \frac{\mathrm{d}\mathbf{k}}{(2\pi)^3} \int_{\mathbb{R}} \mathrm{d}t\, F(t - \mathrm{i}\beta) \exp\left(-\mathrm{i}t \frac{\hbar^2 |\mathbf{k}|^2}{2m}\right) \rho_\beta^{(-)}(\mathbf{k}, z_\infty), \qquad (2.106)$$

where $\rho_\beta^{(-)}(\mathbf{k}, z_\infty)$ is the kernel of $\rho_\beta^{(-)}$ (cf. eqn (2.90)).

The definition of $\mathscr{Z}(\mathbb{R})$ implies that the Payley-Wiener theorem applies here, allowing the variable shift $t - \mathrm{i}\beta \to s$. But then $\mathscr{X} = \mathscr{Y}$ and so $\phi_{\beta\bar{\rho}}$ is $(\beta, \tau_{\beta\bar{\rho}})$-KMS.

From Tomita's theory of *modular Hilbert algebras* (Takasaki [1]), it follows that if $\mathscr{A}_{\beta\bar{\rho}}''$ is a factor, as it is, then $\phi_{\beta\bar{\rho}}$ is the unique $(\beta, \tau_{\beta\bar{\rho}})$-KMS state; the set of extremal KMS states reduces to the Gibbs state:

$$\mathfrak{E}(\mathscr{A}_{\beta\bar{\rho}}''; \tau_{\beta\bar{\rho}}(\mathbb{R})\text{-KMS}) = \{\phi_{\beta\bar{\rho}}\}; \qquad (2.107)$$

the left-hand side indicates the set of extremal $(\beta, \tau_{\beta\bar{\rho}})$-KMS states.

As $\phi_{\beta\bar{\rho}}$ vanishes on the set $\mathscr{A} \backslash \mathscr{A}^{\mathrm{e}}$, these results carry over readily to $\mathscr{A}^{\mathrm{e}}$. This leads to the conclusion stated above: there is no phase transition in this model.

3

The ideal Bose gas

3.1. Introduction

It is a striking feature of quantum theory that systems consisting of one species of particle obey either boson (symmetric) of fermion (antisymmetric) statistics (Pauli (1)). As there is no compelling experimental evidence to do so, we shall not consider systems obeying parastatistics which may arise, e.g. from trilinear relations amongst the fields.

For relativistic systems, spin and space variables are inextricably combined, as in the Dirac equation. For such systems the natural group of space-time motions is the Poincaré group, the semi-direct product of the Lorentz group and the space-time translation group. All the finite-dimensional irreducible faithful unitary representations are labelled by two parameters, the mass and the spin $[m, s]$. In this context, fermion systems have odd half-integral spins $s \in \{\frac{1}{2}, +\frac{3}{2}, \ldots\}$ and boson systems have integral spins $s \in \{0, 1, 2, \ldots\}$; we shall only consider non-zero masses $m > 0$. A deep result of field theory, known as 'the connection between spin and statistics' (see Pauli (4); Streater and Wightman [1]) relates fermion systems to anticommuting fields and boson systems to commuting fields (for spacelike separations). This in turn relates to anticymmetric (resp. symmetric) Fock-Cook spaces on which the fields operate.

In the non-relativistic approximation, the spin and space motions separate. The Dirac equation, for example, goes over into a set of Schrödinger equations whose components are connected by the 2×2 Pauli matrices (Pauli (3)). In general, the one-particle space for a system of spin s particles in the non-relativistic approximation is $\mathbb{L}^2(\mathbb{R}^3, \mathbb{C}^{2s+1})$, whose elements are functions with values in the $(2s+1)$-dimensional Euclidean space $\mathbb{C}^{2s+1}$. For non-relativistic models, it is possible to build artifically either a symmetric or an antisymmetric Fock-Cook space over $\mathbb{L}^2(\mathbb{R}^3, \mathbb{C}^{2s+1})$ for any spin. Artifical since one ought to use antisymmetry (resp. symmetry) for $s \in \{\frac{1}{2}, \frac{3}{2}, \ldots\}$ (resp. $\{0, 1, \ldots\}$).

In the previous chapter we used the antisymmetric Fock-Cook space for spin zero $s = 0$, primarily for convenience; it is not difficult to fill in the additional details necessary to redo the ideal Fermi gas when $s = \frac{1}{2}$. In this chapter we shall consider $s = 0$ again, but symmetrizing the Fock-Cook space this time. We will thus be considering spinless bosons.

The ideal Bose gas is due in the first instance to Bose (1) and Einstein (1), who applied statistics to an assemblage of Planck oscillators. Einstein pointed out that it was possible for an arbitrarily large number of oscillator particles (bosons) to all have zero momentum. This, he said, would resemble a condensation process as in thermodynamics, but would take place in momentum space.

In 1938, F. London, who was very much concerned with long-range quantum correlations in matter, proposed that this model described the strange properties of liquid helium IV. This isotope undergoes a phase transition to liquid helium II, so-called, when its temperature decreases below 2·2 K. As London says (London [1], p.4), the hypothesis that a gas model describes such an exotic liquid is strange. Its validity rests in the agreement between model and experiment. Such agreement as there is is more qualitative than quantitative. One does have confidence that the λ-point transition of helium IV at 2·2 K is of the same nature as the phase transition in the Bose gas model, though. When convenient, we shall use the intuitively appealing London-Tisza [Tisza (1)] language of the two-fluid model. One interprets the properties of helium II, in this language, as due to the existence of two liquids, normal helium and super helium. This latter component moves without friction with either the vessel container or the normal liquid. The density $\rho_s(\beta)$ of the super component increases monotonically from zero at the transition temperature $\beta = \beta_c$ to the whole prescribed density at zero temperature: $\rho_s(\infty) = \bar{\rho}$.

The algebraic formulation is seen to give a very much better understanding of this phase transition than was possible before its advent. We shall find that above the transition, the grand canonical state, i.e. the global Gibbs state, is the unique KMS state and equals the canonical state at the temperature and density. Below the transition, however, the Gibbs state is no longer pure; neither are the canonical states. Each canonical state $\psi_{\beta\rho}$ decomposes into a one-parameter family of pure states $(\phi_{\beta\rho\theta})_{\theta \in T}$ associated with the gauge angles (phases), $\{-\pi \leqslant \theta < \pi\} = T$. The Gibbs state is then a further superposition of canonical states, weighted with Kac's density function, which we shall derive.

In view of the fact that the phase θ is canonically conjugate to the number operator (Dell'Antonio (2); Gille and Manuceau (1); Rocca and Sirugue (1)), this (extremal) decomposition is related to numbers of bosons, in the old language. What really happens is that, below the transition, the gauge transformations are not unitarily implemented; there is no number operator. And one may trace this phenomenon back to the minus sign in $\{1 - z_n \exp[\beta\omega_n(\kappa)]\}^{-1}$, the oscillator factor.

Another system which is partially represented by this model is the simple vibrational modes of an infinite crystal. Such crystal vibrations are known as *phonons* and have the dispersion law $\omega(k) = ck$ for small $k \in \mathbf{R}^+$, with characteristic velocity c. The simple acoustical branch corresponds to the normal component. It is natural to inquire whether the condensation is seen in crystals; this would correspond to a super sound mode known as second sound. This question has been pursued by Landau and his school (Landau and Lifschitz [1]; Abrikosov, A. A., Gor'kov, L. P., and Dzyaloghinskii, I. Ye [1]). An algebraic formulation of an Einstein crystal (countably many linearly interacting oscillators) more closely related to the physical system of a simple crystal, is found in Verbeure and Verboven (1 - 3) and Verboven (1,2).

The first algebraic treatment of the ideal Bose gas was that of Araki and Woods (1). There is an incorrect identification of the canonical and grand canonical state in this paper. This was pointed out and corrected by Lewis and Pulé (1), who gave a detailed analysis of this model, employing both general local regions and general boundary conditions for the Hamiltonians (Lewis (1)). Cannon (1) subsequently published a further analysis of this model making explicit the convergence of the canonical states. He considered only cubical regions and cyclic boundary conditions, which we consider sufficient for this book, and so we shall follow his account. Klauder and Streit (1) have analysed a class of CCR representations, including their GNS constructions, which contain the ones for this model (cf. Robinson (1)).

3.2. Configuation space and Fock-Cook space

In this second model we consider a gas of massive spinless bosons moving freely in $\mathbb{R}^3$. As the configuation space $\mathbb{R}^3$ is the same as for the Fermi gas, so is the family of local regions $\mathscr{L}$ and the distinguished absorbing subfamily $\mathscr{M} = \{V_n : n \in \mathbb{N}\}$ of cubes of edges nL centred at the origin (cf. eqn (2.1)).

The boson Fock-Cook space is constructed in analogy with, but is not identical to, the fermion Fock-Cook space. For each $V \in \mathscr{L}'$ the one-particle space is the same $\mathbb{L}^2(V)$, but the n-particle space $\mathbb{L}^2_S(V^n)$ is now the Hilbert space of *symmetric* square-integrable functions from V^n to $\mathbb{C}$. Formally, the norm on $\mathbb{L}^2_S(V^n)$ is the same as that on $\mathbb{L}^2_A(V^n)$ and we do not distinguish them notationally, namely,

$$\|f^{(n)}\|^2 = \int_{V^n} |f^{(n)}(\xi_1, \ldots, \xi_n)|^2 \, d\xi_1, \ldots, d\xi_n. \tag{3.1}$$

But for those particular elements of $\mathbb{L}^2_A(V^n)$ (resp. $\mathbb{L}^2_S(V^n)$) which are anti-symmetric (resp. symmetric) tensor products of one-particle vectors in $\mathbb{L}^2(V)$, the above norm leads to a determinant (resp. permanent) for inner products; so they are actually quite different spaces.†

Accordingly, the Fock-Cook space for this boson model is

$$\mathscr{H}_B(V) = \bigoplus_{n=0}^{\infty} \mathbb{L}^2_S(V^n) \quad (V \in \mathscr{L}'), \tag{3.2}$$

the B subscript referring to the boson nature of the system. Concurrently, much of the previous notation can be carried over from the Fermi gas, namely,

† A permanent is like a determinant, but there is no sign change between successive cofactors in the Laplace expansion.

$$\Phi = \oplus\Phi^{(n)};\ \|\Phi\|^2 = \sum \|\Phi^{(n)}\|^2 < \infty \qquad \text{for } \Phi \in \mathscr{H}_{\mathrm{B}}(V);$$

and

$$P^{(n)}_V : \mathscr{H}_{\mathrm{B}}(V) \to \mathbf{L}^2_{\mathrm{S}}(V^n)$$
$$P^{(n)}_V\,[\Phi] = \Phi^{(n)} \tag{3.3}$$

for the n-particle projection operator within V.

As it will turn out that the boson fields are unbounded, we must construct a manifold to serve as the domain for these fields. So for $V \in \mathscr{L}'$ we define the non-closed weak direct sum

$$\mathscr{H}^{(0)}_{\mathrm{B}}(V) = \sum_{n\in\mathbf{N}}^{\oplus} [\mathbf{L}^2_{\mathrm{S}}(V^n) \cap \mathscr{D}(\mathbb{R}^{3n})] \quad (V \in \mathscr{L}') \tag{3.4}$$

with the relative topology from $\mathscr{H}_{\mathrm{B}}(V)$. Hereafter, $\mathscr{H}^{\#}_{\mathrm{B}}(V)$ shall denote either $\mathscr{H}_{\mathrm{B}}(V)$ or $\mathscr{H}^{(0)}_{\mathrm{B}}(V)$ indifferently. Note that $\mathscr{H}^{(0)}_{\mathrm{B}}(V)$ is dense in $\mathscr{H}_{\mathrm{B}}(V)$.

The isotony relations hold; there exist injection mappings

$$i^{\#}(V,W) : \mathscr{H}^{\#}_{\mathrm{B}}(V) \to \mathscr{H}^{\#}_{\mathrm{B}}(W) \quad (V \subset W \in \mathscr{L}) \tag{3.5}$$

of the local spaces into those associated with containing volumes. Corresponding to eqn (2.5), the local spaces admit a tensor product structure. With the injections made explicit, one has the relation

$$\mathscr{H}^{\#}_{\mathrm{B}}(U \cup V) = i^{\#}(U, U\cup V)\,[\mathscr{H}^{\#}_{\mathrm{B}}(U)] \otimes i^{\#}(V, U\cup V)\,[\mathscr{H}^{\#}_{\mathrm{B}}(V)]$$
$$(U, V \in \mathscr{L}). \tag{3.6}$$

We shall use the abbreviations† $\mathscr{H}^{\#}_{\mathrm{B}}(V_n) \equiv \mathscr{H}^{\#}_n$; $\mathscr{H}^{\#}_{\mathrm{B}}(\mathbb{R}^3) \equiv \mathscr{H}^{\#}$; $i^{\#}(V_n, V_m) = i^{\#}_{n,m}$; and if $i^{\#}(V) : \mathscr{H}^{\#}_{\mathrm{B}}(V) \to \mathscr{H}^{\#}_{\mathrm{B}}(\mathbb{R}^3)$ is the indicated injection, $i^{\#}(V_n) \equiv i^{\#}_n$.

Then $\mathscr{H}$ is a Hilbert space inductive limit over the cubical regions:

$$\mathscr{H} = \varinjlim \{i_{nm}(\mathscr{H}_n) : n, m \in \mathbf{N}; m > n\}; \tag{3.71}$$

but as $\mathscr{H}^{(0)}$ is not closed, we have instead that

$$\mathscr{H}^{(0)} = \bigcup_{n\in\mathbf{N}} i^{(0)}_n\,[\mathscr{H}^{(0)}_n], \tag{3.7b}$$

and note that we do not complete this subset.

3.3. The algebras

3.3.1. The boson fields

The boson annihilation and creation operators are defined formally just as the fermion fields are, for every $f \in \mathscr{D}(V)$, $V \in \mathscr{L}'$ and $\Phi \in \mathscr{H}^{(0)}_{\mathrm{B}}$ we set‡

$$P^{(n)}_V\,[a_V(\bar{f})\,\Phi]\,(\xi) = (n+1)^{\frac{1}{2}} \int_V \overline{f(\mathbf{y})}\,P^{(n+1)}_V\,[\Phi]\,(\mathbf{y};\xi)\,\mathrm{d}\mathbf{y} \qquad (\xi \in V^n), \tag{3.8a}$$

† No confusion with $\mathscr{H}_{\mathrm{F}}(V_n)$, etc. will result from this notation.

‡ By $\bar{f}$ we mean complex conjugation.

$$P_V^{(n)}[a_V^*(f)\Phi](\xi) = (n)^{\frac{1}{2}} \sum_{j=1}^{n} P_V^{(n-1)}[\Phi](\xi_1,\ldots,\hat{\xi}_j,\ldots,\xi_n) f(\xi_j). \qquad (3.8b)$$

The restriction $f \in \mathscr{D}(V)$ ensures that the domain of the fields is stable under their action $a_V^{\#}(f) : \mathscr{H}_B^{(0)}(V) \to \mathscr{H}_B^{(0)}(V)$; hence polynomials in the fields have $\mathscr{H}_B^{(0)}(V)$ as their domain. The test function space, written as $\mathscr{D}(V)$, is the subset of $\mathscr{D}(\mathbb{R}^3)$ containing functions whose support is within V and equipped with its own LF-topology (Choquet [1]; Robertson and Robertson [1]; Trèves [1]).† Note that we could define fields from $\mathbb{L}^2(V)$, but that would not solve the domain problem as the fields are unbounded. And even if we so extended our definition, in order to have localizability we would have to restrict these fields to test functions $f \in \mathscr{D}(V)$ when defining the algebras in any event.

From this definition it follows that $a_V^{\#}$ is (conjugate) linear from $\mathscr{D}(V)$: $a_V^{\#}(f+g) = a_V^{\#}(f) + a_V^{\#}(g)$; $a_V^*(cf) = c\, a_V^*(f)$ for $c \in \mathbb{C}$; $a_V^*(f) = [a_V(\bar{f})]^*$. Further, they obey the canonical commutation relations (CCR) on $\mathscr{H}_B^{(0)}(V)$, namely,

$$[a_V^{\#}(\bar{f}), a_V^{\#}(g)]_- = 0 \qquad (3.9a)$$

$$[a_V(\bar{f}), a_V^{\#}(g)]_- = \langle f,g\rangle_V\, \mathbb{1}_V, \qquad (3.9b)$$

where $\mathbb{1}_V$ is viewed as an $\mathscr{H}_B^{(0)}(V)$ operator. Because of the CCR, these operator-valued distributions are not bounded and one way to proceed is to exponentiate them. This idea, due to Weyl in the finite-dimensional case and to Segal [1] in the field setting is the method we employ. First we define the field (our notation is not universal)

$$\phi_V(f) = (2)^{-\frac{1}{2}}[a_V^*(f) + a_V(\bar{f})]^{\sim} \quad (f \in \mathscr{D}(V)), \qquad (3.10)$$

which is self-adjoint on $\mathscr{H}_B^{(0)}(V)$. The sum $a_V^*(f) + a_V(\bar{f})$ has $\mathscr{H}_B^{(0)}(V)$ for its domain; as $\mathscr{H}_B^{(0)}(V)$ is not closed, neither is this operator. But it is essentially self-adjoint and so has self-adjoint extensions which are closed. Any one of them could be chosen, but we mean the one corresponding to cyclic boundary conditions for the Hamiltonian, which is quadratic in the fields. We shall only need to be explicit about the Hamiltonian, and so content ourselves with writing a tilde in eqn (3.10) to indicate the extension. The Weyl field for the region $V \in \mathscr{L}'$ is then defined to be

$$W_V(f) = \exp[\mathrm{i}\phi_V(f)] \quad (f \in \mathscr{D}(V)), \qquad (3.11)$$

which also can be extended to all $f \in \mathbb{L}^2(V)$; but we shall not do so. The CCR imply that W_V obeys the so-called Weyl form of the commutation relations (which we shall also denote by CCR):

$$W_V(f)\, W_V(g) = W_V(f+g) \exp[-\tfrac{1}{2}\mathrm{Im}\langle f,g\rangle_V]. \qquad (3.12)$$

The expression

$$\delta(f,g) = -\tfrac{1}{2}\mathrm{Im}\langle f,g\rangle_V$$

is a *symplectic form* on $\mathbb{L}^2(V)$; and so W_V: $\mathscr{D}(V) \to \mathbb{B}[\mathscr{H}_B(V)]$ is a *projective*

† This topology is strictly finer than the relative topology on $\mathscr{D}(V)$ inherited from $\mathscr{D}(\mathbb{R}^3)$.

representation of the additive abelian group of $\mathscr{D}(V)$ with multiplier $\exp[-\frac{1}{2}\,\mathrm{Im}\,\langle f,g\rangle_V]$. A symplectic form on a complex vector space is a real antisymmetric non-degenerate bilinear form. If G is a locally compact group and U is a representation of G by unitary operators on some Hilbert space, then U is said to be a projective representation if the group property takes the form $U(gh) = M(g,h)\,U(g)\,U(h)$. Here the function $M : G \times G \to \mathbb{C}$ is said to be a $\mathbb{C}$-*multiplier* and must satisfy certain identities for consistency. In our case, $M(\cdot) = \exp \mathrm{i}\delta(\cdot)$ with δ the symplectic form $-\frac{1}{2}\,\mathrm{Im}\langle,\rangle$. The reader is referred to Varadarajan [1] for a full account of the theory of multiplier representations and other references.

In calculations it is useful to know the restricted BCH formula, which is only formal unless convergence and existence is discussed, namely,

$$\exp(A)\exp(B) = \exp(A + B + \tfrac{1}{2}\,c\mathbb{1}), \tag{3.13}$$

when $[A,B]_- = c\mathbb{1}$. Applied to the Weyl field, it enables us to rewrite W_V in terms of $a_V^{\#}$ in a sort of 'normal-ordered' (Streater and Wightman [1]) form

$$W_V(f) = \exp(-\tfrac{1}{4}\|f\|_V^2)\exp\{a_V^*[(2)^{-\frac{1}{2}}\,\mathrm{i}f]\}\exp\{a_V[(2)^{-\frac{1}{2}}\,\mathrm{i}\bar{f}]\}. \tag{3.14}$$

Now let $e_\kappa \in \mathbf{L}^2(V_n)$, $\kappa \in \widetilde{V}_n$, be the function

$$e_\kappa(\xi) = (nL)^{-\frac{3}{2}}\exp(\mathrm{i}\kappa\cdot\xi) \quad (\xi \in V_n), \tag{2.44a}$$

so that $\{e_\kappa : \kappa \in \widetilde{V}_n\}$ is the orthonormal basis of $\mathbf{L}^2(V_n)$ adapted to toroidal boundary conditions. Using the bijection $\eta_n : \widetilde{V}_n \to \mathbf{N}$ of §2.5, we define

$$a_p^{*(n)} = a_{V_n}^*(e_\kappa) \quad (\eta_n(\kappa) = p), \tag{3.15}$$

which is formally similar to (2.45); but we now have a family of independent boson (as opposed to fermion) oscillators, and (2.46) is replaced by

$$[a_p^{*(n)}, a_q^{(n)}]_- = \delta_{pq}, \tag{3.16}$$

other commutators vanishing.

3.3.2. *The algebras*

It seems as though there is no canonical choice of boson algebras (Wieringa (1)). Our choice for the local subalgebras are the C^*-algebras on $\mathscr{H}_\mathrm{B}(V)$ consisting of polynomials in the Weyl fields and their uniform limits; we write

$$\mathscr{A}(V) = \text{un.cl.-}\langle W_V(f) : f \in \mathscr{D}(V)\rangle \quad (V \in \mathscr{L}); \tag{3.17}$$

the reader will recall that $\langle\cdot\rangle$ means algebraic span in this context: all finite-order polynomials in the indicated variables modulo the CCR. A typical polynomial is of the form $\sum z_j\, W_V(f_j)$ where the sum is finite and $z_j \in \mathbb{C}$, $f_j \in \mathscr{D}(V)$ for $j = 1, 2, \ldots$, (order of polynomial).

Our notation for the relevant injective mappings of the local subalgebras into containing ones is our usual one:

$$j(V,W) : \mathscr{A}(V) \to \mathscr{A}(W) \quad (V \subset W \in \mathscr{L}). \tag{3.18}$$

We also use our usual abbreviations adapted to cubic regions, namely $\mathscr{A}(V_n) \equiv \mathscr{A}_n$ and $j(V_n, V_m) = j_{nm}$.

The local subalgebras are isotonic and locally commutative; if $U \cap V = \emptyset$ then

$$[\mathscr{A}(U), \mathscr{A}(V)]_- = 0 \tag{3.19}$$

up to an imbedding into any $\mathscr{A}(W)$ with $W \supset U \cup V$.

The quasilocal C^*-algebra is the inductive limiting algebra

$$\mathscr{A} = \underset{\longrightarrow}{\lim}\{j_{nm}(\mathscr{A}_n) : n, m \in \mathbf{N}; m > n\}; \tag{3.20}$$

the systemic local algebra is the union

$$\mathscr{A}_{\mathrm{L}} = \bigcup_{V \in \mathscr{L}} j(V)[\mathscr{A}(V)], \tag{3.21}$$

where $j(V) : \mathscr{A}(V) \to \mathscr{A}$ is the injection into the quasilocal algebra. Then, in accordance with the general scheme, $\mathscr{A}_{\mathrm{L}}$ is norm-dense in $\mathscr{A}$:

$$\mathscr{A} = \text{un.cl.-}\mathscr{A}_{\mathrm{L}}. \tag{3.22}$$

It now follows that $\mathscr{A}$ may be directly defined in terms of Weyl fields:

$$\mathscr{A} = \text{un.cl.-}\langle W(f) : f \in \mathscr{D}(\mathbf{R}^3)\rangle, \tag{3.23}$$

where $W_{\mathbf{R}^3}(f)$ is written $W(f)$ for brevity. Thus we have constructed the kinematics for this model in accordance with our general scheme.

3.4. Symmetries

The pertinent symmetries of this model are space translations and gauge transformations. Although the defining formulae for these symmetries are formally similar to the corresponding Fermi gas definitions, the spectra of the operators in question are rather different.

For space translations, the isometric translation groups are

$$\begin{aligned} S_V(\xi) &: \mathscr{A}(V) \to \mathscr{A}(V+\xi) \quad (V \in \mathscr{L}; \xi \in \mathbf{R}^3), \\ S(\xi) &: \quad \mathscr{A} \to \mathscr{A}, \end{aligned} \tag{3.24}$$

with defining formulae

$$P_V^{(n)}[S_V(\xi)\Phi](x_1, \ldots, x_n) = P_V^{(n)}[\Phi](x_1-\xi, \ldots, x_n-\xi)$$
$$(\xi \in \mathbf{R}^3; x_i \in V, V \in \mathscr{L}'), \tag{3.24a}$$

provided we write $S_{\mathbf{R}^3} \equiv S$.

The corresponding algebraic transformations are

$$\begin{aligned} \sigma_V(\xi) &: \mathscr{A}(V) \to \mathscr{A}(V+\xi), \\ \sigma(\xi) &: \mathscr{A} \to \mathscr{A} \\ \sigma_V(\xi)[A] &= S_V(\xi)\, A\, S_V(-\xi) \quad (A \in \mathscr{A}(V) \text{ or } \mathscr{A}), \end{aligned} \tag{3.25}$$

with $\sigma_{\mathbf{R}^3} \equiv \sigma$.

The fact that $\sigma(\mathbb{R}^3)$ is an automorphism group of $\mathscr{A}$ follows since finite space translations preserve differentiability and compact support, so that e.g. (up to an imbedding) $\mathscr{D}(V) \subset \mathscr{D}(\mathbb{R}^3)$ and $\mathscr{D}(V+\xi) \subset \mathscr{D}(\mathbb{R}^3)$.

Gauge transformations for this model follow the general scheme. The (non-closed) number operator N_V for $V \in \mathscr{L}$ has domain $\mathscr{H}^{(0)}_{\mathrm{B}}(V)$ and is given by the generalized strong sum (here we consider only V_n)

$$N_{V_n} = \sum_{p \in \mathbb{N}}^{(\mathrm{s})} a_p^{*(n)} a_p^{(n)} \tag{3.26a}$$

and

$$N_{\mathbb{R}^3} = \sum_{p \in \mathbb{N}} a^*(\varphi_p)\, a(\bar{\varphi}_p), \tag{3.26b}$$

where $\{\varphi_p\}$ is an orthonormal basis for $\mathbb{L}^2(\mathbb{R}^3)$.

The Fock-Cook space decomposition is (here all $V \in \mathscr{L}'$ are considered)

$$N_V = \bigoplus_{n=0}^{\infty} n\, P_V^{(n)} \quad (V \in \mathscr{L}'); \tag{3.26c}$$

the domain of N_V is

$$\left\{\Phi \in \mathscr{H}_{\mathrm{B}}(V) : \sum_{n=0}^{\infty} n^2 \|\Phi^{(n)}\|^2 < \infty\right\}.$$

Eqns (3.26a) and (3.26b) define an essentially self-adjoint operator. ($\mathscr{H}^{(0)}_{\mathrm{B}}(V)$ is not closed). Eqn (3.26c) is the minimal self-adjoint extension of these, and we abuse notation, calling them all number operators and using the same symbol.

The crucial distinction between bosons and fermions is that the spectrum of $a_p^{*(n)} a_p^{(n)}$ is $\mathbb{Z}^+$ for bosons and $\{0, 1\}$ for fermions. An examination of this aspect of the matter is found in Dell'Antonio (2) and Gille and Manuceau (1).

As for the general scheme, the automorphism group

$$\begin{aligned} \Gamma_V &: T \to \mathrm{Aut}[\mathscr{A}(V)] \quad (V \in \mathscr{L}'), \\ \Gamma_V(\theta)[A] &= U_V(\theta)\, A\, U_V(-\theta), \end{aligned} \tag{3.27}$$

together with the implementing unitary group

$$U_V(\theta) = \exp(\mathrm{i}\theta N_V),$$

this gives the algebraic form of the gauge transformations.

3.5. The Fock-Cook state

The symmetric nature of the boson Fock-Cook space does not affect the existence of a Fock-Cook vacuum vector

$$\Omega_V = 1 \oplus 0 \oplus \ldots . \quad (V \in \mathscr{L}')$$

nor a Fock-Cook condition

$$a_V(f)\, \Omega_V = 0 \quad (\forall f \in \mathbb{L}^2(V)).$$

The corresponding vector state $\omega_V^{(0)} \in \mathfrak{S}\,[\mathscr{A}(V)]$ $(V \in \mathscr{L}')$, the Fock-Cook state for this model, can be explicitly computed. A computation analogous to that for the Fermi gas gives

$$\omega_V^{(0)}\,[a_V(\bar{f}_1)\ldots a_V(\bar{f}_n)\,a_V^*(g_1)\ldots a_V^*(g_m)] \\ = \delta_{nm}\,\mathrm{perm}\,|<f_i, g_j>_V|\,, \tag{3.28a}$$

where a permanent is like a determinant, but without any sign changes between succeeding cofactors. There is another form for $\omega_V^{(0)}$, in terms of the Weyl field rather than the $a_V^{\#}$, namely,

$$\omega_V^{(0)}\,[\mathrm{W}_V(f)] = \exp(-\tfrac{1}{4}\,\|f\|_V^2); \tag{3.28b}$$

this follows from eqn (3.14) for W_V in terms of $a_V^{\#}$, and from using the CCR and $\exp[a_V(f)]\,\Omega_V = \Omega_V$, which follows from the Fock-Cook condition. Note the Gaussian quadratic form this takes. This expression is an expectation generating functional for the state in the sense of abstract probability theory. See Choquet [1], Vol. III and Gel'fand and Vilenkin [1] for measures on topological spaces; and Parthasarathy and Schmidt [1] for the pertinent probability theory.

Viewed as states on $\mathscr{A}$, the $\omega_{V_n}^{(0)}$ converge to $\omega_{\mathbb{R}^3}^{(0)} \equiv \omega^{(0)}$ in the weak*-topology (cf. eqn (2.34)). As $\omega_{V_n}^{(0)}$ is not a state on $\mathscr{A}$ *per se*, we extend it to $\mathfrak{S}(\mathscr{A})$ precisely as we did in eqn (2.34) for the Fermi Fock-Cook state.

3.6. Local dynamics

There is absolutely no difference in the definition of the local Hamiltonians for this boson model as opposed to the Fermi model; nor even of the $\mathbb{R}^3$ one-particle dynamical operators. But the consequences for the algebras are different.

Thus

$$\bar{h}_n,\ u_n^{(1)}(t),\ \sigma_n^{(1)}(\beta),\ U_t^{(n)},\ \bar{H}_n,\ \sigma_\beta^{(n)},\ \text{and}\ \tau^{(n)}(\mathbb{R})$$

are as in §2.6, eqns (2.48)-(2.56). Furthermore, the definitions of $\bar{h}$ and $u_t^{(1)}$ for $\mathbb{R}^3$ are the same. But although $u_n^{(1)}(t)$ leads to an automorphism group $\tau^{(n)}(\mathbb{R}) \in \mathrm{Aut}(\mathscr{A}_n)$, the one-particle unitary group $u_t^{(1)}$ does not lead to an automorphism group of $\mathscr{A}$. Let us now examine this problem. Recall that $u_n^{(1)}$ converges in the strong $\mathbb{L}^2(\mathbb{R}^3)$-topology to $u^{(1)}$ on $\mathscr{D}(\mathbb{R}^3)$ (cf. eqn (2.64a)).

For the Weyl fields $W_{V_n} \equiv W_n$, the action of $\tau^{(n)}$ is given by

$$\tau_t^{(n)}\,W_n(f) = W_n\,[u_n^{(1)}(t)\,f] \qquad (f \in \mathscr{D}(V_n)). \tag{3.29a}$$

Combining this with (2.64a) gives

$$(\text{str. } \mathscr{H}_{\mathrm{B}})\text{-}\lim_{n\to\infty} j_n \circ \tau_t^{(n)}[W_n(f)] = W[u_t^{(1)}(f)] \qquad (f \in \mathscr{D}(\mathbb{R}^3)). \tag{3.29b}$$

But (2.63) tells us that $u_t^{(1)}(f) \notin \mathscr{D}(\mathbb{R}^3)$, so that $W[u_t^{(1)}(f)] \notin \mathscr{A}$. Thus we do not have an automorphism group of $\mathscr{A}$ (although the mapping $W(f) \to W[u_t^{(1)}(f)]$ leads to a one-parameter group of automorphisms of $\mathbb{B}(\mathscr{H}_{\mathrm{B}})$). We shall come back to this problem after we have computed the global Gibbs state.

3.7. The local Gibbs states

Recall that our notation is $\sigma_\beta^{(n)} = \exp(-\beta\bar{H}_n)$ with $\bar{H}_n = H_n - \mu_n N_n$ for the reduced Hamiltonian, and we shall write $\Xi_\beta^{(n)} = \exp(-\beta H_n)$ when the full Hamiltonian is meant. Then

$$\begin{aligned} E_{\beta\bar{\rho}}^{(n)}(A) &= \mathrm{tr}_n[\sigma_\beta^{(n)} A] \\ &= \mathrm{tr}_n[\Xi_\beta^{(n)} (z_n)^{N_n} A] \quad (A \in \mathscr{A}_n), \end{aligned} \tag{3.30a}$$

with the fugacity z_n separated out explicitly. The local Gibbs state $\phi_{\beta\bar{\rho}}^{(n)} \in \mathfrak{S}(\mathscr{A}_n)$ is defined to be

$$\phi_{\beta\bar{\rho}}^{(n)}(A) = E_{\beta\bar{\rho}}^{(n)}(A)/E_{\beta\bar{\rho}}^{(n)}(\mathbb{1}). \tag{3.30b}$$

The following calculation will sketch the derivation of the formula

$$\begin{aligned} &\phi_{\beta\bar{\rho}}^{(n)}[\exp(-\mathrm{i}\xi N_n/|V_n|)\, W_n(f)] \\ &= \omega_n^{(0)}[W_n(f)]\, \widetilde{K}_n^{(+)}(\xi,\bar{\rho}) \exp[2(nL)^{-3} \sum_{\kappa\in\widetilde{V}_n} z_n |\widetilde{f}(\kappa)|^2 \times \\ &\quad \times \{z_n - \exp[\beta\omega_n(\kappa) + \mathrm{i}(\xi/|V_n|)]\}^{-1} \quad . \end{aligned} \tag{3.31}$$

In including the number operator in the argument, $\phi_{\beta\bar{\rho}}^{(n)}$ must be viewed as a state on all of $\mathbb{B}[\mathscr{H}_\mathrm{B}(V_n)]$; the requisite extension follows by choosing $A \in \mathbb{B}[\mathscr{H}_\mathrm{B}(V_n)]$ in eqn (3.30); we shall not notationally distinguish between the full state and its restriction.

The first factor in this formula is the V_n-Fock-Cook state:

$$\omega_n^{(0)}[W_n(f)] = \exp(-\tfrac{1}{4}\|f\|_n^2); \tag{3.28c}$$

the function $\omega_n(\kappa) = \hbar^2|\kappa|^2/2m$ ($\kappa \in \widetilde{V}_n$) is the energy; and $\widetilde{K}_n^{(+)}$ is the Fourier transform of the boson Kac density, which is defined by

$$\widetilde{K}_n^{(+)}(\xi,\bar{\rho}) = \phi_{\beta\bar{\rho}}^{(n)}[\exp(-\mathrm{i}\xi N_n/|V_n|)]\,; \tag{3.32}$$

we shall show that this may be written as

$$\begin{aligned} \widetilde{K}_n^{(+)}(\xi,\bar{\rho}) &= \prod_{\kappa\in\widetilde{V}_n} \{1 - z_n \exp[-\beta\omega_n(\kappa)]\} \times \\ &\quad \times \{1 - z_n \exp[-\beta\omega_n(\kappa) + \mathrm{i}(\xi/|V_n|)]\}^{-1}. \end{aligned} \tag{3.33}$$

Just as for the Fermi gas, the Kac density relates the canonical states $\psi_{\beta\bar{\rho}}^{(V)}$ to the grand canonical state. Eqns (2.93)-(2.96) are valid for this model:

$$\begin{aligned} \phi_{\beta\bar{\rho}}^{(n)}(A) &= \sum_{j\in\mathbb{N}} \phi_{\beta\bar{\rho}}^{(n)}(P_{V_n}^{(j)})\; \psi_{\beta\, j/|V_n|}^{(V_n)}(A) \\ &= \sum_{j\in\mathbb{N}} \widetilde{K}^{(+)}(j/|V_n|,\bar{\rho})\; \psi_{\beta\, j/|V_n|}^{(V_n)}(A), \end{aligned} \tag{3.34}$$

where $P^{(j)}_{V_n} : \mathscr{H}_n \to \mathbf{L}^2_S[(V_n)^j]$; note the constraint $j = (nL)^3\bar{\rho}$.

In the calculation which follows we use the fields $a_\eta^{\#(n)}$ defined in eqn (3.15) with $\eta_n(\kappa) = p$ as fixed notation. These fields have $\mathscr{H}_n^{(0)}$ as domain, and so the equalities in the calculation refer to this manifold. With these conventions, the operator inside the trace in eqn (3.31) can be written

$$\sigma_\beta^{(n)} \exp(-\mathrm{i}\xi N_n/|V_n|)\, W_n(f) = \prod_{\kappa\in\tilde{V}_n} A_\kappa \exp(-\tfrac{1}{4}\|f\|_n^2), \tag{3.35}$$

with

$$\begin{aligned} A_\kappa = {} & \exp\{[\beta\, \epsilon_n(\kappa) + \mathrm{i}(\xi/|V_n|)]\, a_p^{*(n)} a_p^{(n)}\} \times \\ & \times \exp[\mathrm{i}(2|V_n|)^{-\frac{1}{2}} \tilde{f}(\kappa)\, a_p^{*(n)}] \times \\ & \times \exp[-\mathrm{i}(2V_n)^{-\frac{1}{2}} \tilde{f}(\kappa)^*\, a_p^{(n)}], \end{aligned} \tag{3.36}$$

which reduces the argument of $E_{\beta\bar{\rho}}^{(n)}$ to a product. The following formulae refer to each mode ($s_p \in \mathbf{C}$ and at this point the expressions are formal):

$$\exp(s_p a_p^{(n)})\, \Omega_0^{(n)} = \Omega_0^{(n)}, \tag{3.37a}$$

with $\Omega_0^{(n)}$ the V_n-Fock-Cook vacuum vector;

$$\|[a_p^{*(n)}]^j \Omega_0^{(n)}\|^2 = j!\,; \tag{3.37b}$$

$$\exp(-s_p\, a_p^{(n)})\, a_p^{*(n)} \exp(s_p\, a_p^{(n)}) = a_p^{*(n)} + s_p; \tag{3.37c}$$

$$\exp(-s_p\, a_p^{*(n)} a_p^{(n)})\, [a_p^{*(n)}]^j\, \Omega_0^{(n)} = \exp(-j s_p)\, [a_p^{*(n)}]^j\, \Omega_0^{(n)}. \tag{3.37d}$$

In view of (3.37d), we may find an orthonormal basis of $\mathscr{H}_n$ whose elements are the eigenfunctions of the number operator. The trace in $E_{\beta\bar{\rho}}^{(n)}$ then is over the modes and leads to

$$\begin{aligned} \mathrm{tr}_p A_\kappa = {} & \sum_{j\in\mathbf{N}} (j!)^{-1} \exp\{-[\beta\epsilon_n(\kappa) + \mathrm{i}(\xi/|V_n|)]^j\} \times \\ & \times \left\langle [a_p^{*(n)}]^j\, \Omega_0^{(n)}, \exp(-\bar{s}_p\, a^{*(n)}) \exp(s_p a_p^{(n)})\, [a_p^{*(n)}]^j\, \Omega_0^{(n)} \right\rangle, \end{aligned} \tag{3.38a}$$

with s_p now taken to be $s_p = -\,\mathrm{i}/[2(nL)^3]^{\frac{1}{2}}\ \tilde{f}(\kappa)^*$; the asterisk is used for the complex conjugate of f here so as not to interfere with the tilde for the Fourier transform. Multiplying $\Omega_0^{(n)}$ by $\exp[s_p\, a_p^{(n)}] \exp[-s_p\, a_p^{(n)}] \equiv \mathbf{1}_n$ and using (3.37c) gives

$$\begin{aligned} & \left\langle [a_p^{*(n)}]^j\, \Omega_0^{(n)}, \exp(-\bar{s}_p\, a_p^{*(n)})\, [a_p^{*(n)} - s_p]^j\, \Omega_0^{(n)} \right\rangle \\ & = \|[a_p^{*(n)} - s_p]^j\, \Omega_0^{(n)}\|^2 \\ & = \sum_{\lambda=0}^{j} (j!/\lambda!) \begin{bmatrix} j \\ \lambda \end{bmatrix} (-1)^\lambda [s_p]^\lambda \end{aligned}$$

for the matrix element. But then $\mathrm{tr}_p A_\kappa$ can be written as

$$\mathrm{tr}_p A_\kappa = \{1 - \exp[-\beta\epsilon_n(\kappa) - \mathrm{i}(\xi/|V_n|)]\}^{-1} \times$$

$$\times \exp\left\{|s_p|^2\{1-\exp[\beta\epsilon_n(\kappa)+i(\xi/|V_n|)]\}^{-1}\right\}. \tag{3.38b}$$

The result (3.31) follows immediately from this.

A more familiar form for this result is in terms of the Bose-Einstein operator $\rho_n^{(+)}(\beta) \in \mathbb{B}[\mathbb{L}^2(V_n)]$:

$$\rho_n^{(+)}(\beta) = -\tfrac{1}{2} \sum_{\kappa \in \widetilde{V}_n} z_n\{z_n - \exp[\beta\omega_n(\kappa)]\}^{-1} e_\kappa \otimes e_\kappa^* \tag{3.39}$$

for then (3.31) leads to

$$\phi_{\beta\bar{\rho}}^{(n)}[W_n(f)] = \omega_n^{(0)}[W_n(f)]\exp\{-\tfrac{1}{2}\langle f, \rho_n^{(+)}(\beta) f\rangle\}. \tag{3.40}$$

Note the all-important minus sign in eqn (3.39).

3.8. The global Gibbs state

We are now in a position to consider the limit for large n of $\phi_{\beta\bar{\rho}}^{(n)}$. The operator convergence problem following from, say, (3.40) just above, is ill-posed. One needs a regularization; this is furnished uniquely by the constraint of constant 'prescribed density'.† Note that the consistency of this regularization involves the distinguished family $\mathcal{M}$, i.e. the choice of the cubes V_n and cyclic boundary conditions. We shall return to this point at the end. Let us proceed with the case in hand; and we do this just as we did for the Fermi gas. To this end we define the function $G_\beta^{(+)} : [0,1] \to \mathbb{R}$,

$$G_\beta^{(+)}(\zeta) = (2\pi)^{-3} \int_{\mathbb{R}^3} \{-\zeta + \exp[\beta\hbar^2|\mathbf{k}|^2/(2m)]\}^{-1}\,\mathrm{dk}, \tag{3.41}$$

which can be seen to be increasing in ζ.

In contrast to the Fermi case, the lowest eigenvalue mode in all the sums *must* be isolated and treated separately, for it can be singular in the limit. The convergence behaviour of all these series must then be expected to be different from the Fermi case. The first change has already appeared; the fugacity range, represented by ζ in $G_\beta^+(\zeta)$, is restricted to $[0,1]$. Note that $G_\beta^{(+)}(1)$ exists. The values of fugacity $z(\beta)$ are, of course, temperature-dependent. We shall see that $z(\beta) = 1$ for all $\beta > \beta_c$ a critical temperature value, and it is this value $\zeta = 1$ which is the singular point of the lowest eigenvalue contributions. We shall place a prime on sums which omit the lowest eigenvalue contribution.

LEMMA 3.1

The limit

$$\lim_{n\to\infty} (nL)^{-3} \sum_{\kappa\in\widetilde{V}_n}{}' \zeta\{-\zeta + \exp[\beta\omega_n(\kappa)]\}^{-1} = G_\beta^{(+)}(\zeta) \tag{3.42}$$

exists uniformly in $\zeta \in [0,1]$.

† That is, there is a unique limit for constant density; other densities are, of course, possible and correspond to different regularizations. For regularizations, see Gel'fand and Shilov [1].

Proof. The prime on the sum refers to the omission of the term $\kappa = 0$. Just as in the Fermi-gas case, only terms for which all the $\kappa_i \neq 0$ ($i = 1, 2, 3$) contribute in the limit. The contributory summands approximate the $G_\beta^{(+)}$ integrand monotonically from below by step functions; the approximation is uniform in (ζ, κ) outside some neighbourhood of the $[0,1] \times \mathbb{R}^3$ origin. More details concerning the proof of this lemma may be found in the paper by Cannon (1).

The prescribed mean density appears through the expression

$$\bar{\rho} = (nL)^{-3} z_n(1-z_n)^{-1} + (nL)^{-3} \sum_{\kappa \in \tilde{V}_n}' z_n\{\exp[\beta\omega_n(\kappa)] - z_n\}^{-1}, \quad (3.43\text{a})$$

which we rewrite in the form

$$(nL)^{-3} z_n(1-z_n)^{-1} = \bar{\rho} - (nL)^{-3} \sum_{\kappa \in \tilde{V}_n}' z_n\{\exp[\beta\omega_n(\kappa)] - z_n\}^{-1}. \quad (3.43\text{b})$$

Now $z_n \to z_\infty$ with n; and if $z_\infty < 1$ the left-hand side of this equality vanishes (for fixed β). The corresponding equation of state is

$$\bar{\rho} = G_\beta^{(+)}(z_\infty) \quad (0 < z_\infty < 1). \quad (3.44\text{a})$$

As $G_\beta^{(+)}$ is increasing with z_∞, $\bar{\rho}$ increases with z_∞ until z_∞ reaches unity. The value of $\bar{\rho}$ at this point will be written ρ_c; it is a critical-point value, namely,

$$\rho_c(\beta) = G_\beta^{(+)}(1). \quad (3.44\text{b})$$

The dependence of ρ_c upon temperature follows from the definition of $G_\beta^{(+)}$, from which we see that there is a triple-relation amongst the intensive parameters $\bar{\rho}$, z_∞, and β. It is only if β is large enough that z_∞ can reach unity and $\bar{\rho}$ reach $\rho_c(\beta)$. For large temperatures (small β), $G_\beta^{(+)}$ can be inverted so that there is a unique relation between $\bar{\rho}$ and z_∞, and the pair (z_∞, β) may be used as independent variables.

But for low temperatures ($\beta \geqslant \beta_c$) the fugacity $z_n \to 1$ converges to unity and eqn (3.44) cannot be uniquely inverted. As we are considering the grand canonical state it is the density which is prescribed, so that $(\bar{\rho}, \beta)$ must be used as independent thermodynamic variables. And in this region the otherwise undefined limit of the left-hand side of eqn (3.43b) (undefined so far since we do not know the asymptotic behaviour of z_n with n) will be defined as the well-defined limit of the right-hand side, namely, $\bar{\rho} - \rho_c(\beta)$. From the two-fluid picture sketched in the introduction to this model, there is a superfluid component with some density $\rho_S(\beta)$ such that $\bar{\rho} = \rho_S(\beta) + \rho_c(\beta)$. Consequently the limit of the left-hand side will be interpreted as the density $\rho_S(\beta)$ of the superfluid condensate. Let $[N]$ (resp. $[S]$) refer to the regions in parameter space where only a normal component exists (resp. a normal and superfluid component coexist). Then first of all, in the $[S]$ region we have

$$\lim_{n \to \infty} (nL)^{-3} z_n(1-z_n)^{-1} = \rho_S(\beta) \quad (z_\infty = 1). \quad (3.45)$$

Actually, $\rho_S(\beta)$ is better written as $\bar{\rho} - \rho_c(\beta)$, with $\rho_c(\beta)$ the normal component density at β; the fixed critical density is then $\rho_c(\beta_c)$ with β_c the inverse critical temperature. Thus

$$\lim_{n\to\infty} (nL)^{-3} z_n (1-z_n)^{-1} = \begin{cases} 0 \text{ for } \bar{\rho} = G_\beta^{(+)}(z_\infty), \text{ when } \bar{\rho} < \rho_c(\beta) \text{ and } z_\infty < 1; & [N] \\ \bar{\rho} - \rho_c(\beta), \quad \text{when } \bar{\rho} \geqslant \rho_c(\beta) \text{ and } z_\infty = 1; & [S] \end{cases} \tag{3.46}$$

with $\rho_c(\beta)$ defined by (3.44b).

This result can now be used to examine the limit of eqn (3.33). First of all we can consider the contribution of the lowest eigenvalue separately†

$$(1-z_n)\{1-z_n \exp[\mathrm{i}(\xi/|V_n|)]\}^{-1}$$

$$= 1 - [(nL)^{-3} z_n (1-z_n)^{-1}]\{\mathrm{i}\xi + \ldots + (\mathrm{i}\xi)^m [m!(nL)^{3m-3}]^{-1} + \ldots\}$$

$$\to \begin{cases} 1 & [N] \\ \{1 - \mathrm{i}\xi[\bar{\rho} - \rho_c(\beta)]\}^{-1} & [S] \end{cases} \tag{3.47}$$

uniformly in ξ in compacta, i.e. for relatively compact neighbourhoods of ξ.

Next we consider the expression

$$\prod_{\kappa\in V_n, \kappa\neq 0} \{1-z_n \exp[-\beta\omega_n(\kappa)]\}\{1-z_n\exp[-\beta\omega_n(\kappa) + \mathrm{i}(\xi/|V_n|)]\}^{-1}$$

$$= \exp\left\{-\sum_{\kappa\in\tilde{V}_n}' \log\left[\{1-z_n\exp[-\beta\omega_n(\kappa)+\mathrm{i}(\xi/|V_n|)]\}\{1-z_n\exp[-\beta\omega_n(\kappa)]\}^{-1}\right]\right\}$$

$$= \exp\left\{+\sum_{\kappa\in\tilde{V}_n}' \left[\{1-z_n\exp[-\beta\omega_n(\kappa)+\mathrm{i}(\xi/|V_n|)]\}\{1-z_n\exp[-\beta\omega_n(\kappa)]\}^{-1}\right] + \text{second order}\right\}$$

$$= \exp\left\{\sum_{\kappa\in\tilde{V}_n}' \left[\mathrm{i}(\xi/|V_n|) z_n \{\exp[\beta\omega_n(\kappa)] - z_n\}^{-1} + \text{second order}\right]\right\}$$

$$\to \exp[\mathrm{i}\xi G_\beta^{(+)}(z_\infty)]. \tag{3.48}$$

This proves Kac's lemma, which we call Lemma 3.2.

LEMMA 3.2

$$\tilde{K}_n^{(+)}(\xi, \bar{\rho}) \to \tilde{K}^{(+)}(\xi, \bar{\rho});$$

$$\tilde{K}^{(+)}(\xi, \bar{\rho}) = \begin{cases} \exp(\mathrm{i}\xi\bar{\rho}) & \bar{\rho} \leqslant \rho_c(\beta) \quad [N] \\ \{1-\mathrm{i}\xi[\bar{\rho}-\rho_c(\beta)]\}^{-1}\exp[\mathrm{i}\xi\rho_c(\beta)] & \bar{\rho} \geqslant \rho_c(\beta) \quad [S] \end{cases}$$

where the convergence is uniform in ξ in compacta.

The Fourier transform of $\tilde{K}^{(+)}$ is the Kac density proper, which can now be computed directly.

† Replacing $-z_n$ by $+z_n$ gives the Fermi case; the corresponding limit is unambiguously equal to unity.

$$K^{(+)}(\rho,\bar\rho) = \begin{cases} \delta(\rho-\bar\rho) & \text{for } \bar\rho<\rho_c(\beta)\ [N] \\ \begin{bmatrix} 0 & \text{if } \rho<\rho_c(\beta) \\ [\bar\rho-\rho_c(\beta)]^{-1}\exp\{[\rho-\rho_c(\beta)][\rho_c(\beta)-\bar\rho]^{-1}\} & \text{if } \rho\geqslant\rho_c(\beta) \end{bmatrix} & \text{and } \bar\rho\geqslant\rho_c(\beta)\ [S] \end{cases} \quad (3.50)$$

The normal density in the hyperussiastic region can be written in terms of Riemann's zeta function $\zeta(w) = \sum_{n=1}^{\infty} n^{-w}$ (Re $w > 0$). In particular, $\zeta(\frac{3}{2}) \simeq 2{\cdot}62$ and we have (cf. eqn (3.44))

$$\rho_c(\beta) = (2\pi M/\hbar^2\beta)^{\frac{3}{2}}\zeta(\tfrac{3}{2}), \quad (3.51)$$

which is the well-known formula for the critical density (cf. Landau and Liftschitz [1]).

Unfortunately, there is still more to compute in order to find the Gibbs state, namely, the limit of the last term in eqn (3.31).

Now the exponent of this last term can be written as

$$-\ (2|V_n|)^{-1} z_n \exp(-i\xi/|V_n|)[1-z_n\exp(-i\xi/|V_n|)]^{-1}|\tilde f(0)|^2 -$$

$$-\ (2|V_n|)^{-1} \sum_{\kappa\in\tilde V_n}{}' \sum_{t=1}^{\infty} (z_n)^t |\tilde f(\kappa)|^2 \exp\{-t[\beta\omega_n(\kappa) + i(\xi/|V_n|)]\}.$$

The first term may be further expanded in powers of $(nL)^{-3}$:

$$-\tfrac{1}{2}[(nL)^{-3}z_n(1-z_n)^{-1} - i\xi(nL)^{-6}z_n(1-z_n)^{-1} + \ldots] \times$$
$$\times\ [1 + i\xi(nL)^{-3}z_n(1-z_n)^{-1} + \ldots]^{-1}|\tilde f(0)|^2,$$

and seen to converge to

$$\begin{cases} 0 & [N] \\ -\tfrac{1}{2}[\bar\rho-\rho_c(\beta)]\{1 + i\xi[\bar\rho-\rho_c(\beta)]\}^{-1}|\tilde f(0)|^2 & [S] \end{cases}$$

In the second term, only the lowest term in a ξ-expansion survives the limit of large n; keeping only this term and resumming over t gives $(2|V_n|)^{-1} \sum'_{\kappa\in\tilde V_n} z_n\{z_n - \exp[\beta\omega_n(\kappa)]\}^{-1}|f(\kappa)|^2$. And just as in Lemma 3.1, this approaches an integral in the limit, uniformly in ξ in compacta. This proves the main theorem, which is:

THEOREM 3.1

For any $\beta, \bar\rho > 0$ and $f \in \mathscr{D}(\mathbf{R}^3)$, the limit

$$\phi^{(n)}_{\beta\bar\rho}[\exp(i\xi N_n/|V_n|)\, W_n(f)]$$

$$\to \tilde K^{(+)}(\xi,\bar\rho)\exp(-\tfrac{1}{4}\|f\|^2)\exp\left\{(16\pi^3)^{-1}\int_{\mathbf{R}^3} z_\infty[z_\infty - \exp|\beta\omega(\kappa)|]^{-1}|\tilde f(\kappa)|^2\,dk\right\} \times$$

$$\times \begin{cases} 1 & [N] \\ \exp\ \tfrac{1}{2}[\rho_c(\beta)-\bar\rho]\{1 + i\xi[\bar\rho-\rho_c(\beta)]\}^{-1}|\tilde f(0)|^2 & [S] \end{cases} \quad (3.52)$$

converges uniformly in ξ in compacta, where $\omega(\mathbf{k}) = \hbar^2|\mathbf{k}|^2/2m$.

When we introduced the Kac density function $K^{(+)}$ we used it to relate the grand canonical Gibbs state to the canonical states for local regions. Now we have shown that the local region Gibbs state converges to a global Gibbs state, and we also know various properties of the density function $K^{(+)}$. If we can show that the canonical states converge in the thermodynamic limit, we can then find a global relation between the canonical and grand canonical states. For the Fermi gas, we assumed the convergence of the canonical states, and because $K^{(-)}(\rho, \bar{\rho}) = \delta(\rho - \bar{\rho})$ the two states were the same: $\phi_{\beta\bar{\rho}} = \psi_{\beta\bar{\rho}}$, which is eqn (2.99). As we already known (cf. eqn (3.50)) that $K^{(+)}$ is not simply a delta function, the canonical states can be expected to be distinct from $\phi_{\beta\bar{\rho}}$ in this model. Cannon (1) has proved the following theorem in this regard.

THEOREM 3.2 (Cannon)
For any $\beta, \rho > 0$ *and* $f \in \mathscr{D}(\mathbb{R}^3)$,

$$\psi_{\beta\bar{\rho}}^{(V_n)}[W_n(f)] \to \psi_{\beta\rho}[W(f)] \tag{3.53}$$

uniformly in compacta with respect to the density $\rho \in (0,\infty)$. *The limiting canonical state is given by the expression* (Araki and Woods (1))

$$\psi_{\beta\rho}[W(f)] = \exp(-\tfrac{1}{4}\|f\|^2) \begin{cases} \exp\left[(16\pi^3)^{-1} \int\limits_{\mathbb{R}^3} z_\infty\{z_\infty - \exp[\beta\omega(\mathbf{k})]\}^{-1} |\tilde{f}(\mathbf{k})|^2 \, d\mathbf{k}\right] & \text{for } \rho < \rho_c(\beta); \\ \exp\left[(16\pi^3)^{-1} \int\limits_{\mathbb{R}^3} \{1 - \exp[\beta\omega(\mathbf{k})]\}^{-1} |\tilde{f}(\mathbf{k})|^2 \, d\mathbf{k}\right] \times & \\ \times J_0\{[2\rho - 2\rho_c(\beta)]^{\frac{1}{2}} |\tilde{f}(0)|\} & \text{for } \rho \geqslant \rho_c(\beta), \end{cases} \tag{3.54}$$

with $\rho = G_\beta^{(+)}(z_\infty)$ *when* $\rho < \rho_c(\beta)$, *where* J_0 *is the zero-order Bessel function of the first kind.*

The proof is long and involves some combinatorics; the reader is referred to Cannon's paper.

Using the connection between the canonical and grand canonical state eqn (3.34), it is clear that the following result is true.

COROLLARY 3.1

$$\phi_{\beta\bar{\rho}} = \int\limits_{\mathbb{R}_*^+}^{\oplus} \psi_{\beta\rho}[K^{(+)}(\rho, \bar{\rho}) \, d\rho]. \tag{3.55}$$

The direct integral is with respect to the GNS triples associated with the states. In terms of images, e.g. $\phi_{\beta\bar{\rho}}(A)$, which are numbers, a complex-valued integral of the usual sort results.† We make this point precise by writing $\phi_{\beta\bar{\rho}} \sim [\mathscr{H}_{\beta\bar{\rho}}, \pi_{\beta\bar{\rho}}, \Omega_{\beta\bar{\rho}}]$ and $\psi_{\beta\rho} \sim [\mathscr{H}_{\beta\rho}^{(c)}, \pi_{\beta\rho}^{(c)}, \Omega_{\beta\rho}^{(c)}]$ for the GNS triples, where the superscriped c refers to canonical. Then, because $\pi_{\beta\rho}^{(c)}$ is not equivalent to $\pi_{\beta\rho'}^{(c)}$ for $\rho \neq \rho'$, the Hilbert space $\mathscr{H}_{\beta\bar{\rho}}$ decomposes, and we have

† $\mathbb{R}_*^+$ stands for $\{0 < x < \infty\}$.

$$_{\beta\bar{\rho}} = \int_{\mathbf{R}_*^+}^{\oplus} H_{\beta\rho}^{(c)} K^{(+)}(\rho, \bar{\rho})\, \mathrm{d}\rho, \tag{3.56a}$$

$$\pi_{\beta\bar{\rho}} = \int_{\mathbf{R}_*^+}^{\oplus} \pi_{\beta\rho}^{(c)} K^{(+)}(\rho, \bar{\rho})\, \mathrm{d}\rho, \tag{3.56b}$$

$$\Omega_{\beta\bar{\rho}} = \int_{\mathbf{R}_*^+}^{\oplus} \Omega_{\beta\rho}^{(c)} K^{(+)}(\rho, \bar{\rho})\, \mathrm{d}\rho. \tag{3.56c}$$

Note that for $\rho \geqslant \rho_c(\beta)$ this is a non-trivial decomposition, as seen from eqn (3.50); for $\rho < \rho_c(\beta)$ there is no decomposition.

The question is: does $\pi_{\beta\rho}^{(c)}$ decompose further in some natural way?—and the answer is yes; there is a spontaneous breakdown of gauge symmetry.

From the integral representation

$$J_0(z) = \int_0^{2\pi} (2\pi)^{-1} \mathrm{d}\theta \exp[-\mathrm{i}z\cos(\theta)], \tag{3.57}$$

for the Bessel function one can see that it might be possible to decompose each $\mathscr{H}_{\beta\rho}^{(c)}$ with respect to angle. We therefore consider the Hilbert space $\mathbf{L}^2(T)$ of square-integrable functions from the circle $T = \{0 \leqslant \theta < 2\pi\}$ to $\mathbf{C}$, equipped with the norm

$$\|f\|_T^2 = \int_0^{2\pi} (2\pi)^{-1} \mathrm{d}\theta\, |f(\theta)|^2. \tag{3.58}$$

Note the shift by π in our convention for T.

By convention we shall write $e \in \mathbf{L}^2(T)$ for the identity constant function $e(\theta) \equiv 1$; note that $\|e\|^2 = 1$.

In what follows we decompose the function $f \in \mathscr{D}(\mathbf{R}^3)$ into real and imaginary parts, writing $\widetilde{f}(\mathbf{k}) = \widetilde{F}(\mathbf{k}) + \mathrm{i}\widetilde{G}(\mathbf{k})$, where F and G are real-valued elements of $\mathscr{D}(\mathbf{R}^3)$; note that $\widetilde{f}(\mathbf{k})^* = \widetilde{F}(\mathbf{k}) - \mathrm{i}\widetilde{G}(\mathbf{k})$.

It is clear that there is a boson density operator $\rho_\beta^{(+)}$, analogous to $\rho_\beta^{(-)}$ for the Fermi model, appearing in the formulae. We take $\rho_\beta^{(+)}$ to be the integral operator on $\mathbf{L}^2(\mathbf{R}^3)$ whose *multiplier*, i.e. the Fourier image of its kernel, is

$$\rho_\beta^{(+)}(\mathbf{k}, z_\infty) = z_\infty\{\exp[\beta\omega(\mathbf{k})] - z_\infty\}^{-1}. \tag{3.59}$$

The operators $[\rho_\beta^{(+)}]^{-\frac{1}{2}}$ and $[1 + \rho_\beta^{(+)}]^{\frac{1}{2}}$ whose multipliers are $[\rho_\beta^{(+)}(\mathbf{k}, z_\infty)]^{\frac{1}{2}}$ and $[1 + \rho_\beta^{(+)}(\mathbf{k}, z_\infty)]^{\frac{1}{2}}$, respectively, are perfectly well defined; they will appear in the decomposition of $\pi_{\beta\bar{\rho}}$.

First though we decompose the Bessel function

$$J_0\{[2\rho - 2\rho_c(\beta)]^{\frac{1}{2}} |\widetilde{f}(0)|\}$$
$$= \int_0^{2\pi} (2\pi)^{-1} \mathrm{d}\theta \exp\{\mathrm{i}[2\rho - 2\rho_c(\beta)]^{\frac{1}{2}} [\widetilde{F}(0)\cos(\theta) - \widetilde{G}(0)\sin(\theta)]\}. \tag{3.60}$$

In accordance with our guess about using $\mathbb{L}^2(T)$ we define the sine and cosine operators on it. S and $C \in \mathbb{B}[\mathbb{L}^2(T)]$ are defined by the formulae

$$\begin{aligned}(Sf)(\theta) &= \sin\theta\, f(\theta),\\ (Cf)(\theta) &= \cos\theta\, f(\theta). \quad (\forall f \in \mathbb{L}^2(T)).\end{aligned} \tag{3.61}$$

But to get to the irreducible representations we decompose $\mathbb{L}^2(T)$ into one-dimensional subspaces, each one a copy of $\mathbb{C}$:

$$\mathbb{L}^2(T) = \int_0^{2\pi\,\oplus} (2\pi)^{-1}\, \mathrm{d}\theta\, \mathbb{L}_\theta; \tag{3.62}$$

this decomposition is discussed in Vilenkin [1].

Corresponding to this, one has the decompositions

$$e = \int_0^{2\pi\,\oplus} (2\pi)^{-1}\, \mathrm{d}\theta\, e_\theta,$$

and

$$S = \int_0^{2\pi\,\oplus} (2\pi)^{-1}\, \mathrm{d}\theta\, S_\theta,$$

$$C = \int_0^{2\pi\,\oplus} (2\pi)^{-1}\, \mathrm{d}\theta\, C_\theta, \tag{3.63}$$

where $e_\theta \equiv 1$, S_θ (resp. C_θ) $\in \mathbb{B}(\mathbb{L}_\theta)$, and $S_\theta f_\theta = \sin(\theta) f_\theta$ (resp. $C_\theta f_\theta = \cos(\theta) f_\theta$) for every $f_\theta \in \mathbf{L}_\theta$.

The full decomposition for the Gibbs state can now be written, and is given by

$$\mathscr{H}_{\beta\bar{\rho}} = \int_{T\times\mathbb{R}_*^+}^{\oplus} (\mathscr{H}_\mathrm{B} \otimes \mathscr{H}_\mathrm{B} \otimes \mathbb{L}_\theta)_{\beta\rho}\, (2\pi)^{-1}\, \mathrm{d}\theta \times K^{(+)}(\rho, \bar{\rho})\, \mathrm{d}\rho, \tag{3.64a}$$

$$\Omega_{\beta\bar{\rho}} = \int_{T\times\mathbb{R}_*^+}^{\oplus} (\Omega_{\mathbb{R}^3} \otimes \Omega_{\mathbb{R}^3} \otimes e_\theta)_{\beta\rho}\, (2\pi)^{-1}\, \mathrm{d}\theta \times K^{(+)}(\rho, \bar{\rho})\, \mathrm{d}\rho, \tag{3.64b}$$

$$\pi_{\beta\bar{\rho}} = \int_{T\times\mathbb{R}_*^+}^{\oplus} (\pi_{\beta\rho\theta})\, (2\pi)^{-1}\, \mathrm{d}\theta \times K^{(+)}(\rho, \bar{\rho})\, \mathrm{d}\rho, \tag{3.64c}$$

where the representation is

$$\begin{aligned}\pi_{\beta\rho\theta}[W(F + \mathrm{i}G)] &= W[(1 + \rho_\beta^{(+)})^{\frac{1}{2}}(F + \mathrm{i}G)] \otimes W[(\rho_\beta^{(+)})^{\frac{1}{2}}(F - \mathrm{i}G)] \otimes\\ &\otimes \exp\{\mathrm{i}\chi_\mathrm{S}[2\rho - 2\rho_\mathrm{c}(\beta)]^{\frac{1}{2}}[\widetilde{F}(0)C_\theta - \widetilde{G}(0)S_\theta]\}.\end{aligned} \tag{3.64d}$$

We have introduced the characteristic function for the superfluid region:

$$\chi_\mathrm{S} = \begin{cases} 0 & \bar{\rho} < \rho_\mathrm{c}(\beta)\\ 1 & \bar{\rho} \geqslant \rho_\mathrm{c}(\beta). \end{cases} \tag{3.64e}$$

Before analysing this result, let us go back to the relation between the Gibbs state and the canonical states. To complete this aspect of the matter, the corollary (eqn (3.55)) must be supplemented by the decomposition

$$\mathscr{H}^{(c)}_{\beta\rho} = \int_0^{2\pi \oplus} (\mathscr{H}_B \otimes \mathscr{H}_B \otimes \mathbf{L}_\theta)\,(2\pi)^{-1}\,d\theta, \tag{3.65a}$$

$$\Omega^{(c)}_{\beta\rho} = \int_0^{2\pi \oplus} (\Omega_{\mathbb{R}^3} \otimes \Omega_{\mathbb{R}^3} \otimes e_\theta)\,(2\pi)^{-1}\,d\theta, \tag{3.65b}$$

$$\pi^{(c)}_{\beta\rho}[W(F+iG)] = \int_0^{2\pi \oplus} (2\pi)^{-1}\,d\theta\; W[1+\rho^{(+)}_\beta)^{\frac{1}{2}}(F+iG)] \otimes W[(\rho^{(+)}_\beta)^{\frac{1}{2}}(F-iG)] \otimes \otimes \exp\{i\chi_S[2\rho-2\rho_c(\beta)]^{\frac{1}{2}}[\tilde{F}(0)C_\theta - \tilde{G}(0)S_\theta]\} \tag{3.65c}$$

associated with the canonical state.

It is obvious from eqn (3.64) that a phase transition occurs, since there is no decomposition for $\bar{\rho} < \rho_c(\beta)$. This phase transition is accompanied by the spontaneous breakdown of gauge symmetry. Let $\alpha \in T$ and consider the gauge transformations (by α) of the states. Recall that the $\mathbb{R}^3$ number operator on $\mathscr{H}_B$ generates a unitary group $U(T)$ and thence an automorphism group $\Gamma(T)$ of $\mathscr{A}$, which is implemented by $U(T)$. The states transform in accordance with $\varphi[\Gamma(\alpha)A] = [\Gamma^*(\alpha)\varphi](A)$, where $A \in \mathscr{A}$, $\varphi \in \mathfrak{S}(\mathscr{A})$, $\alpha \in T$.

In particular, the action of $\Gamma^*(\alpha)$ on $\phi_{\beta\bar{\rho}}$ comes to the replacement

$$[\tilde{F}(0)C_\theta - \tilde{G}(0)S_\theta] \to \tfrac{1}{2}[\tilde{F}(0) - i\tilde{G}(0)]\exp[i(\alpha+\theta)] + + \tfrac{1}{2}[\tilde{F}(0) + i\tilde{G}(0)]\exp[i(\alpha-\theta)] \tag{3.66}$$

in eqn (3.64d).

For notational convenience, let us define the state $\phi_{\beta\rho\theta} \in \mathfrak{S}(\mathscr{A})$ by means of eqn (3.64), so that we may write

$$\phi_{\beta\bar{\rho}} = \int_{T\times\mathbb{R}^+_*}^{\oplus} \phi_{\beta\rho\theta}\,(2\pi)^{-1}\,d\theta \times K^{(+)}(\rho,\bar{\rho})\,d\rho. \tag{3.64f}$$

Then the action of $\Gamma^*(T)$ given above shows that

$$\Gamma^*(\alpha)(\phi_{\beta\rho\theta}) \neq \phi_{\beta\rho\theta}, \tag{3.67a}$$

i.e. $\phi_{\beta\rho\theta} \notin \mathfrak{S}(\mathscr{A};\Gamma(T))$. But upon integrating over gauge angles, it is clear that

$$\Gamma^*(\alpha)(\phi_{\beta\bar{\rho}}) = \phi_{\beta\bar{\rho}}. \tag{3.67b}$$

That is, the global Gibbs state, $\phi_{\beta\bar{\rho}} \in \mathfrak{S}[\mathscr{A};\Gamma(T)]$, is T-invariant. Thus we have a decomposition of a T-invariant state ($\phi_{\beta\bar{\rho}}$) into T-non-invariant components ($\phi_{\beta\rho\theta}$); as we shall see, the $\phi_{\beta\rho\theta}$ are extremal KMS states. Hence this is a spontaneous breakdown of T-symmetry associated with the extremal KMS decomposition.

On the other hand, since $|\exp(i\mathbf{k}\cdot\xi)[\widetilde{F}(\mathbf{k}) \pm i\widetilde{G}(\mathbf{k})]|$ equals $|\widetilde{F}(\mathbf{k}) \pm i\widetilde{G}(\mathbf{k})|$, and $\exp(i\mathbf{k}\cdot\xi)\widetilde{f}(\mathbf{k})|_{\mathbf{k}=0}$ equals $\widetilde{f}(0)$, there is no spatial symmetry breakdown.

As we shall see below, the $\pi_{\beta\rho\theta}$ decomposition is the *central decomposition* and so the algebra decomposes:

$$\pi_{\beta\bar{\rho}}(\mathscr{A})'' = \int^{\oplus}_{T\times\mathbb{R}^+_*} \pi_{\beta\rho\theta}(\mathscr{A})''\,(2\pi)^{-1}\,d\theta \times K(\bar{\rho},\rho)\,d\rho.$$

We shall write $\pi_\delta(\mathscr{A}) = \mathscr{A}_\delta$ and $\pi_\delta(\mathscr{A})'' = \mathscr{A}''_\delta$ for either $\delta = (\beta,\rho,\theta)$ or $(\beta,\bar{\rho})$.

3.9. Global dynamics

The great difficulty in setting up a dynamical scheme for this model is that for $t \neq 0$ and $f \neq 0$ with $f \in \mathscr{D}(\mathbb{R}^3)$, $u^{(1)}_t(f) \in \mathscr{S}(\mathbb{R}^3) \setminus \mathscr{D}(\mathbb{R}^3)$. In general, any CCR representation associated with a state having the finite mean density property—e.g. $\pi_{\beta\bar{\rho}}$ and $\phi_{\beta\rho}$—can be extended continuously from $\mathscr{D}(\mathbb{R}^3)$ to $\mathscr{S}(\mathbb{R}^3)$: this was shown by Lanford and Robinson (2). Then $W_{\beta\bar{\rho}}(f) \equiv \pi_{\beta\bar{\rho}}[W(f)]$ makes sense for all $f \in \mathscr{S}(\mathbb{R}^3)$. We shall prove this extension theorem directly for $W_{\beta\bar{\rho}}$; having done so, $W_{\beta\bar{\rho}}(u^{(1)}_t(f))$ is then well defined and leads to an automorphism group of $\mathscr{A}_{\beta\bar{\rho}}''$ on $\mathscr{H}_{\beta\bar{\rho}}$.

It is standard that the usual $\mathscr{S}(\mathbb{R}^3)$ topology is given through the countable family of semi-norms

$$\mathfrak{p}_{st}(\varphi) = \max_{s+|\mathbf{t}|} \sup_{\mathbf{k}\in\mathbb{R}^3} |(1+|\mathbf{k}|^2)^{s/2} D^{\mathbf{t}}(\varphi)(\mathbf{k})|, \tag{3.68}$$

where $s \in \mathbb{N}$, $\mathbf{t} \in (\mathbb{Z}^+)^3$. In particular, a sequence $\varphi_j \in \mathscr{S}(\mathbb{R}^3)$ converges to zero in this topology iff $\mathfrak{p}_{st}(\varphi_j) \to 0$ for all $s \in \mathbb{N}$, $\mathbf{t} \in (\mathbb{Z}^+)^3$ (Choquet [1]; Gel'fand and Shilov [1], Vol. 2; Robertson and Robertson [1]; Trèves [1]).

Now if $Q : \mathscr{S} \to \mathbb{C}$ is a quadratic form on $\mathscr{S}(\mathbb{R}^3)$, it is continuous iff $\varphi_j \to 0$ as above implies that $Q(\varphi_j) \to 0$.

The argument of the exponential in $\phi_{\beta\bar{\rho}}[W(f)]$ is a quadratic form over $\mathscr{D}(\mathbb{R}^3)$; let us formally extend it to $f \in \mathscr{S}(\mathbb{R}^3)$ and see if it is continuous. For if it is, it extends to a well-defined continuous form on $\mathscr{S}$, which extension has the same formula.

First consider the part common to both the $[N]$ and $[S]$ regions, namely,

$$Q(f) = (16\pi^3)^{-1} \int_{\mathbb{R}^3} z_\infty\{z_\infty - \exp[\beta\omega(\mathbf{k})]\}^{-1} |\widetilde{f}(\mathbf{k})|^2\,d\mathbf{k}\quad (\forall f \in \mathscr{S}(\mathbb{R}^3)). \tag{3.69}$$

Now let $\{\varphi_j \in \mathscr{D}(\mathbb{R}^3)\}$ be a sequence converging to zero in the $\mathscr{S}(\mathbb{R}^3)$ topology.†
As $\mathscr{S}(\mathbb{R}^3)$ is stable under Fourier transformation, $\widetilde{\varphi}_j \in \mathscr{S}(\mathbb{R}^3)$ and $\widetilde{\varphi}_j \to 0$ in $\mathscr{S}(\mathbb{R}^3)$. Then

† The topology induced on $\mathscr{D}(\mathbb{R}^3)$ by the $\mathscr{S}(\mathbb{R}^3)$ topology is coarser than the LF-topology (Choquet [1]; Robertson and Robertson [1]; Trèves [1]).

$$|Q(\varphi_j)| \leqslant (2\pi)^{-3}\mathfrak{p}_{2r\,0}(\widetilde{\varphi}_j) \int_{\mathbb{R}^3} \rho_\beta^{(+)}(\mathrm{k}, z_\infty)\,(1+|\mathrm{k}|^2)^{-r}\,\mathrm{dk}$$
$$\leqslant \mathfrak{p}_{2r\,0}(\widetilde{\varphi}_j)\,\rho_c(\beta).$$

And as $\widetilde{\varphi}_j \to 0$ implies $\mathfrak{p}_{2r\,0}(\widetilde{\varphi}_j) \to 0$, we have $Q(\varphi_j) \to 0$. Thus this form extends continuously to $\mathscr{S}(\mathbb{R}^3)$.

The condensate appears through the contact term, proportional to $|\widetilde{f}(0)|^2$. But $\varphi_j \to 0$ implies that $|\widetilde{\varphi}_j(0)|^2 \to 0$, so this quadratic form also extends. Finally, the boundedness and continuity of the exponential exp : $\mathbb{R} \to \mathbb{R}$ means that $\phi_{\beta\bar{\rho}}$ is $\mathscr{S}(\mathbb{R}^3)$-continuous, i.e. we may simply choose $f \in \mathscr{S}(\mathbb{R}^3)$. However, the CCR extension so obtained does not lead to an automorphism of $\mathscr{A}_{\beta\bar{\rho}}$ proper; a consideration of the topologies, which we shall not supply here—the interested reader is referred back to Lanford and Robinson—shows that it is an automorphism of the weak closure $\mathscr{A}''_{\beta\bar{\rho}}$. Thus we define

$$\begin{aligned} \tau_{\beta\bar{\rho}} &: \mathbb{R} \to \mathrm{Aut}[\mathscr{A}''_{\beta\bar{\rho}}]; \\ \tau_{\beta\bar{\rho}}(t)[W_{\beta\bar{\rho}}(f)] &= W_{\beta\bar{\rho}}[u_t^{(1)}(f)] \quad (\forall f \in \mathscr{D}(\mathbb{R}^3)). \end{aligned} \tag{3.70}$$

By the same proof we gave for the $\mathscr{S}(\mathbb{R}^3)$-continuity of the forms, $\tau_{\beta\bar{\rho}}$ is weakly continuous in time. And because $\phi_{\beta\bar{\rho}}$ has the finite mean density property, $\mathscr{H}_{\beta\bar{\rho}}$ is separable (Hugenholtz and Wieringa (1)). Standard semi-group (Hille and Phillips [1]) theory then implies that $\tau_{\beta\bar{\rho}}$ is unitarily implemented by a strongly continuous unitary group $U_{\beta\bar{\rho}}(\mathbb{R})$,

$$\tau_{\beta\bar{\rho}}(t)[A] = U_{\beta\bar{\rho}}(t)A\,U_{\beta\bar{\rho}}(-t) \quad (\mathrm{A} \in \mathscr{A}''_{\beta\bar{\rho}}). \tag{3.71}$$

The generator of $U_{\beta\bar{\rho}}$—which exists by virtue of Stone's theorem—is 'the Liouville operator' for this state:

$$U_{\beta\bar{\rho}}(t) = \exp(\mathrm{i}t\,L_{\beta\bar{\rho}}). \tag{3.72}$$

Taking the first derivative in the strong topology gives the pertinent 'Liouville equation' (cf. eqn (2.100)):

$$[L_{\beta\bar{\rho}}, Q]_- = \mathrm{i}^{-1}\,\text{str.-}\lim_{t\to 0} \tfrac{\mathrm{d}}{\mathrm{d}t}\,\tau_{\beta\bar{\rho}}(t)\,[Q] \quad (Q \in \mathscr{A}''_{\beta\bar{\rho}}). \tag{3.73}$$

Let $\Phi_{\beta\bar{\rho}} \in \mathfrak{S}(\mathscr{A}''_{\beta\bar{\rho}})$ be the unique continuous extension of the Gibbs state $\phi_{\beta\bar{\rho}} \in \mathfrak{S}(\mathscr{A})$ to $\mathscr{A}''_{\beta\bar{\rho}}$; it is defined by

$$\Phi_{\beta\bar{\rho}}(Q) = (\Omega_{\beta\bar{\rho}}, Q\,\Omega_{\beta\bar{\rho}})_{\mathscr{H}_{\beta\bar{\rho}}} \qquad (Q \in \mathscr{A}''_{\beta\bar{\rho}}). \tag{3.74}$$

A calculation similar to, but longer than, the one we did for the Fermi gas shows that

$$\begin{aligned} &\int_{\mathbb{R}} F(t-\mathrm{i}\beta)\,\Phi_{\beta\bar{\rho}}\,\{W_{\beta\bar{\rho}}(f)\,W_{\beta\bar{\rho}}[u_t^{(1)}(g)]\}\,\mathrm{d}t \\ &= \int_{\mathbb{R}} F(t)\,\Phi_{\beta\bar{\rho}}\,\{W_{\beta\bar{\rho}}[u_t^{(1)}(g)]\,W_{\beta\bar{\rho}}(f)\}\,\mathrm{d}t \end{aligned} \tag{3.95}$$

for all $F \in \mathscr{Z}(\mathbb{R})$ and suitable f and g. But this is the KMS integral equation, so that $\Phi_{\beta\bar{\rho}}$ is $(\beta, \tau_{\beta\bar{\rho}})$-KMS.

Dell'Antonio (1) has analysed the types of $\mathscr{A}''_{\beta\bar{\rho}}$. His results classify the types by means of the density multiplier function $\rho^{(+)}_{\beta}(\mathbf{k}, z_\infty)$, and are as follows.

Normal region. In the [*N*] region, $\mathscr{A}''_{\beta\bar{\rho}}$ is unitarily equivalent to $\mathscr{A}''_{\beta'\bar{\rho}'}$ iff

$$\rho^{(+)}_{\beta}(\mathbf{k}, z_\infty) = \rho^{(+)}_{\beta'}(\mathbf{k}, z'_\infty) \quad (\forall \mathbf{k} \in \mathbb{R}^3). \tag{3.96}$$

The algebras $\mathscr{A}''_{\beta\bar{\rho}}$ are Type-III, unless

$$\rho^{(+)}_{\beta}(\mathbf{k}, z_\infty)\, f(\mathbf{k}) = 0 \quad (\forall f \in \mathscr{L}(\mathbb{R}^3)), \tag{3.97}$$

whence it is Type-I.

Using the theory of modular Hilbert algebras, the same theorem of Takesaki (1) as was quoted for the Fermi gas is applicable: $\mathscr{A}''_{\beta\bar{\rho}}$ is extremal KMS when it is a factor. And as the $\mathscr{A}''_{\beta\bar{\rho}}$ are all factors in the [*N*] region, they are all the unique $(\beta, \tau_{\beta\bar{\rho}})$-extremal KMS state (each $\beta, \bar{\rho}$).

Hyperussiastic region. In the [*S*] region, the $\mathscr{A}''_{\beta\rho\theta}$ are factors, in fact, Type-III factors. Each corresponding state $\Phi_{\beta\rho\theta} \in \mathfrak{S}(\mathscr{A}''_{\beta\rho\theta})$ is $(\beta, \tau_{\beta\rho\theta})$-KMS, where $\tau_{\beta\rho\theta}$ is defined by

$$\tau_{\beta\rho\theta} : \mathbb{R} \to \mathrm{Aut}(\mathscr{A}''_{\beta\rho\theta}),$$
$$\tau_{\beta\rho\theta}(t)\,[W_{\beta\rho\theta}(f)] = W_{\beta\rho\theta}\,[u^{(1)}_t(f)]. \tag{3.98}$$

The proof that it is KMS is very nearly identical to the proof that $\Phi_{\beta\bar{\rho}}$ is KMS. And then the theorem of Takesaki implies that the $\Phi_{\beta\rho\theta}$ are extremal $(\beta, \tau_{\beta\rho\theta})$-KMS.

But this says that the integral decomposition (eqn (3.64f)) is the extremal–or ergodic–KMS decomposition of $\phi_{\beta\bar{\rho}}$. It is also the case that the corresponding algebraic decomposition

$$\mathscr{A}''_{\beta\bar{\rho}} = \int^{\oplus}_{T\times\mathbb{R}^+_*} \mathscr{A}''_{\beta\rho}\,(2\pi)^{-1}\,\mathrm{d}\theta \times K^{(+)}(\rho, \bar{\rho})\,\mathrm{d}\rho \tag{3.99}$$

is the central decomposition.

Let us note what would happen if more general regions and boundary conditions were chosen. Lewis and Pulé (1) considered regions which were dilations of some distinguished one which was star-shaped w.r.t. an interior point (the origin) and had piecewise smooth boundary. Their boundary conditions on the Hamiltonian required $c_1F(\partial V) + c_2\,\partial F/\partial n(\partial V) = 0$ if F was in the domain of $H(V)$, where c_1 and c_2 are constants and ∂V is the boundary of V. This results in the replacement $\rho_S(\beta) \to \lambda\rho_S(\beta)$ in the Gibbs state, where λ depends upon c_1 and c_2. If $c_1 = 0$, $c_2 = 1$ then $\lambda = 1$, just as for cyclic boundary conditions. The other extreme, $c_1 = 1$, $c_2 = 0$ gives $\lambda = \tilde{e}_1(0)$, where e_1 is the eigenfunction of the one-particle Hamiltonian corresponding to the lowest eigenvalue. Although the dependence is not very sensitive, we have here a counterexample to the 'folk theorem' that the effect of the boundary conditions disappears in the limit. Alternatively, one might feel that the local boundary conditions really are irrelevant to the global equilibrium state which ought to be found by global methods. For the ideal Bose gas at least, such a method is available using the CCR and the KMS condition (Moya (1); Sewell (2)).

4

The BCS model

4.1. Introduction

In 1911, Kamerlingh-Onnes (1) found that if mercury is cooled below a definite critical value it loses its electrical resistance. This would seem to be the first known example of a hyperussiastic effect, in this case superconductivity. All hyperussiastic phenomena known are thermodynamically stable; they are described by impure thermal equilibrium states. That this should be so for superconductivity may be deduced from the Meissner effect, which we now describe.

If a superconducting sample, $T < T_c$, is placed in a magnetic field, it is found experimentally that the field is excluded from the interior of the sample (neglecting a small penetration length). On the other hand, if $T > T_c$, the sample and the magnetic field come to thermal equilibrium with a non-zero field present inside the sample. Upon lowering the temperature of the sample, the field is suddenly expelled from the interior when the critical temperature $T = T_c$ is reached; and remains excluded for all $T < T_c$. Thus we may conclude that superconductivity is independent of the past history of the sample and is therefore a purely thermodynamical phenomenon.

At present, a complete microscopic theory of superconductivity does not exist, but the derivation of the generalized BCS Hamiltonian (Bardeen, Cooper, and Schrieffer (1)) from the Frölich Hamiltonian (Frölich (1, 2)) by Bogoliubov (1) and coworkers (Bogoliubov, Tolmachev, and Shirkov [1]) is a great step forward. Moreover, the theory of Bardeen, Cooper, and Schrieffer predicts the correct kind of phase transition, quasi-particle spectrum, and the Meissner effect (Meissner and Ochsenfeld (1)).

We intend to start from the BCS model Hamiltonian in the strong coupling limit and its accompanying approximation of considering only *Cooper pair* states (Cooper (1)). Using the quasi-spin formalism of Thirring and coworkers (Baumann, Eder, Sexl, and Thirring (1)), we shall then derive the explicit expression for the thermodynamic Gibbs state and the dynamical evolution in the island around it. In common with the ideal Bose gas, the onset of the hyperussiastic state is signalled by the spontaneous breakdown of gauge symmetry, and the time translations do not describe an automorphism of the quasilocal algebra, but of the weak closure of its thermodynamic representations.

For an excellent review of the historical and physical background of superconductivity and the various theories proposed to describe it, we strongly recommend the article of Schafroth (1) which also contains an extensive bibliography. London's book on superconductivity (London [1]) is recommended for an account of the early semi-macroscopic theories.

4.2. The model kinematics

In the BCS model, superconductivity is due to the collective behaviour of the metallic conduction electrons. These elctrons find it energetically favourable to form pairs whose total momentum relative to the Fermi surface (Landau and Lifschitz [1]) is zero. As these 'Cooper pairs' (Cooper (1)) are formed from pairs of electron raising operators, they will have complicated commutation relations on the electron Fock-Cook space. But it is not the electron Fock-Cook space which is relevant. Rather, it is the subspace formed from Cooper pair operators acting on the vacuum which is the underlying Hilbert space $\mathscr{H}$ for this model.

As operators on the Cooper pair space $\mathscr{H}$ the fields we must deal with can be described completely with the same spin formalism we used in Chapter 2. This is particularly advantageous to us since we wish to use the 'strong coupling limit' BCS Hamiltonian, and this Hamiltonian takes an especially simple form in terms of spin operators.

We shall refer to the number of Cooper pairs as the mode number, and this must be a positive integer. For m modes, $m \in \mathbf{N}$, the Hilbert subspace $\mathscr{H}_m$ of the full space $\mathscr{H}$ is the tensor product of m copies of $\mathbf{C}^2$; hence $\mathscr{H}_m$ is 2^m-dimensional. Let us write

$$\mathscr{H}_m = \bigotimes_{1}^{m} \mathbf{C}^2. \tag{4.1}$$

We distinguish the vector $\mathbf{e} = (1, 0)$ in $\mathbf{C}^2$, and use it to identify $\mathscr{H}_m$ with a subspace of $\mathscr{H}_n$ for $n > m$ in the obvious way: the injection

$$i_{mn} : \mathscr{H}_m \to \mathscr{H}_n \quad (m < n) \tag{4.2a}$$

is defined by

$$i_{mn}(\mathbf{v}) = \mathbf{v} \bigotimes_{m+1}^{n} \mathbf{e}. \tag{4.2b}$$

The full Hilbert space $\mathscr{H}$ is the inductive limit of these local Hilbert spaces:

$$\mathscr{H} = \varinjlim \{i_{mn}[\mathscr{H}_m] : m, n \in \mathbf{N}; m < n\}. \tag{4.3}$$

This construction depends upon the choice of neutral vector $\mathbf{e} \in \mathbf{C}^2$ at each mode. This choice distinguishes the unit product vector

$$\Omega_+ = \otimes_{\mathbf{N}} \mathbf{e} \tag{4.4}$$

in $\mathscr{H}$; $\mathscr{H}$ is sometimes known as the incomplete direct product space of $\mathbf{C}^2$ with itself, with respect to Ω_+. Note, however, that $\mathscr{H}$ is a complete Hilbert space.

Any vector in $\mathscr{H}$ is a linear combination of product vectors each differing from Ω_+ at only a finite number of modes.

With the notation of §2.6 we write $\mathfrak{B}[m]$ for the local algebra on $\mathscr{H}_m$; our choice for $\mathfrak{B}[m]$ is obvious, since $\mathscr{H}_m$ is finite-dimensional,

$$\begin{aligned} \mathfrak{B}[m] &= \mathbb{B}(\mathscr{H}_m) \\ &= \bigotimes_1^m \mathbb{B}(\mathbb{C}^2). \end{aligned} \tag{4.5}$$

The tensor product is merely algebraic here as $\mathfrak{B}[m]$ is finite-dimensional. These local algebras form an inductive chain of algebras upon using the unit operator as a neutral operator:

$$\begin{aligned} \varphi_{mn} &= \mathfrak{B}[m] \to \mathfrak{B}[n]; \\ \varphi_{mn}(A) &= A \otimes \mathbb{1}_{n\backslash m} \quad (m<n). \end{aligned} \tag{4.6}$$

Then the C^*-inductive limit of this chain is a C^*-algebra $\mathfrak{B}$, the quasilocal algebra for the model:

$$\mathfrak{B} = \varinjlim\{\varphi_{mn}(\mathfrak{B}[m]) : m, n \in \mathbb{N}; m<n\}. \tag{4.7}$$

Suppose we label the modes sequentially, i.e. the spin operators generating $\mathfrak{B}[m]$, for example, will be labelled $\{J_p^{(\alpha)}[m] : p = 1, 2, \ldots, m; \alpha = 1, 2, 3\}$. For the quasilocal algebra we have the generating set $\{J_p^{(\alpha)} : p \in \mathbb{N}; \alpha = 1, 2, 3\}$. These operators are Pauli matrices $\sigma^{(\alpha)}$ at one mode and unit operators elsewhere, e.g.

$$J_1^{(\alpha)}[m] = \sigma^{(\alpha)} \otimes \mathbb{1}_{m\backslash 1} \tag{4.8a}$$

and

$$J_1^{(\alpha)} = \sigma^{(\alpha)} \otimes \mathbb{1}_{\mathbb{N}\backslash 1}. \tag{4.8b}$$

This notation is the same as used in §2.6, eqns (2.39)-(2.43).

It will prove convenient to consider the total spin operators for the $\mathfrak{B}[m]$; these are defined by the formula

$$L_m^{(\alpha)} = 2 \sum_{p=1}^{m} J_p^{(\alpha)}[m], \tag{4.9}$$

with $\alpha = 1, 2$, or 3 and $m \in \mathbb{N}$. We use these operators to generate one parameter groups by exponentiation; the product of these three groups is

$$\omega_m(\mathbf{a}) = \prod_{j=1}^{3} \exp(\mathrm{i}a_j L_m^{(j)}), \tag{4.10}$$

with $\mathbf{a} = (a_1, a_2, a_3) \in \mathbb{R}^3$. With the thermodynamic limit in mind we shall need to consider $\omega_n(\mathbf{b}/n)$, where $\mathbf{b} \in \mathbb{R}^3$ is independent of the mode number n. This completes our formulation of the kinematics for this model.

4.3. The local Hamiltonian

The BCS model Hamiltonian is non-local; an interaction-term quadratic in the underlying electron fields has a non-local kernel. This is somewhat obscured by the spin formalism. For in this notation, the BCS Hamiltonian for $\mathfrak{B}[n]$ is

$$H_n(\text{BCS}) = \sum_{p=1}^{n} \epsilon_p^{(n)} \{1 - 2 J_p^{(3)}[n]\} - \\ - \sum_{p,q=1}^{n} \mathscr{V}_{pq}^{(n)} J_p^{(+)}[n] J_q^{(-)}[n]. \tag{4.11}$$

Our treatment of this model will be concerned exclusively with the strong coupling limit. We define this limit by taking $\epsilon_p^{(n)} = \epsilon$ and $\mathscr{V}_{pq}^{(n)} = 8g/n$, with ϵ and g independent of n. The result of substituting these values into $\mathscr{H}_n(\text{BCS})$ is our model Hamiltonian:

$$H_n = \epsilon n - \epsilon L_n^{(3)} - (2g/n) L_n^{(+)} L_n^{(-)}. \tag{4.12}$$

In accordance with our general scheme, the Hamiltonian H_n generates the Gibbs semi-group $\sigma_\beta^{(n)} = \exp(-\beta H_n)$ and the unitary time translation group $U_t^{(n)} = \exp(\mathrm{i}tH_n)$. The former semi-group is used to construct the local Gibbs state $\phi_\beta^{(n)} \in \mathfrak{S}(\mathfrak{B}[n])$:

$$\phi_\beta^{(n)}(A) = \operatorname{tr}_n(\sigma_\beta^{(n)} A)/\operatorname{tr}_n(\sigma_\beta^{(n)}), \tag{4.13}$$

and the latter group is used to generate the time translation local automorphism group $\tau^{(n)}(\mathbb{R}) \subset \operatorname{Aut}(\mathfrak{B}[n])$:

$$\tau_t^{(n)}(A) = U_t^{(n)} A\, U_{-t}^{(n)}. \tag{4.14}$$

Before computing the local Gibbs state $\phi_\beta^{(n)}$, we should like to use the permutation symmetry of the Hamiltonian H_n to simplify matters. Since the spin operators generate the local algebras, if one knowns the thermal average $\phi_\beta^{(n)}(\mathbf{p})$ for any polynomial $\mathbf{p}$ in the spin operators, one knows the state $\phi_\beta^{(n)}$. At first sight, this seems to imply that we must compute the $3n$-parameter generating function

$$\phi_\beta^{(n)}[\exp(\mathrm{i} \textstyle\sum a_j J_j^{(1)}[n]) \exp(\mathrm{i} \sum b_j J_j^{(2)}[n]) \exp(\mathrm{i} \sum c_j J_j^{(3)}[n])], \tag{4.15a}$$

but this is not so.

To see this, let π_n be the permutation group on n letters and let $g \in \pi_n$ be a permutation. The group π_n acts as an automorphism group $\mathscr{G} : \pi_n \to \operatorname{Aut}(\mathfrak{B}[n])$ when we set

$$\mathscr{G}_g : J_p^{(\alpha)}[n] \to J_{g(p)}^{(\alpha)}[n], \tag{4.16a}$$

i.e. we permute indices. Clearly it follows that $L_n^{(\alpha)}$ and hence H_n are π_n-invariant:

$$\begin{aligned} \mathscr{G}_g[L_n^{(\alpha)}] &= 2 \sum_{p=1}^{n} J_{g(p)}^{(\alpha)}[n] \\ &= 2 \sum_{g^{-1}(j)=1}^{n} J_j^{(\alpha)}[n] \\ &= L_n^{(\alpha)} \end{aligned} \tag{4.16b}$$

$$\begin{aligned} \mathscr{G}_g[H_n] &= \epsilon_n - \epsilon\, \mathscr{G}_g[L_n^{(3)}] - (2g/n)\, \mathscr{G}_g[L_n^{(+)}]\, \mathscr{G}_g[L_n^{(-)}] \\ &= H_n. \end{aligned} \tag{4.16c}$$

Note that because $\mathscr{G}_g$ is an automorphism of $\mathfrak{B}[n]$, $\mathscr{G}_g[AB] = \mathscr{G}_g[A]\,\mathscr{G}_g[B]$ for any $A, B \in \mathfrak{B}[n]$.

Now the local Gibbs state $\phi_\beta^{(n)}$ is a normal state formed from $\sigma_\beta^{(n)} = \exp(-\beta H_n)$. It follows that $\phi_\beta^{(n)}$ is a symmetric state. That is, if we write

$$\mathscr{S} = \sum_{g \in \pi_n} \mathscr{G}_g$$

for the symmetrizing operator, we find that

$$\phi_\beta^{(n)}(\mathscr{S}[A]) = \phi_\beta^{(n)}(A) \tag{4.16d}$$

for every $A \in \mathfrak{B}[n]$. This means that a knowledge of $\phi_\beta^{(n)}(\mathscr{S}[A])$ for every $A \in \mathfrak{B}[n]$ is sufficient to define $\phi_\beta^{(n)}$. If we apply this observation to the generating function (4.15), we see that (cf. eqn (4.10))

$$\phi_\beta^{(n)}[\omega_n(\mathbf{b}/n)] \equiv F_n(\mathbf{b}) \tag{4.15b}$$

is a generating function for $\phi_\beta^{(n)}$. In the next section we shall compute this three-parameter function explicitly.

4.4. The local Gibbs state

Having used some group theory to reduce the necessary computation for $\phi_\beta^{(n)}$ from a $3n$- to a three-parameter function, we intend to use a bit more group theory to actually find $F_n(\mathbf{b})$.

Because H_n can be written entirely in terms of spin operators, the representation theory of SU(2) will enable $F_n(\mathbf{b})$ to be found. All the representation theory we shall use will be taken from the book of Vilenkin [1], to which the reader is referred for details. By, for example, V.3.1(2) we mean Chapter 3, §1, eqn (2) of Vilenkin [1].

The generic SU(2) element can be written in terms of a set of Euler angles φ, θ, Ψ with ranges $0 \leqslant \varphi < 2\pi$, $0 < \theta < \pi$, and $-2\pi \leqslant \Psi < 2\pi$, namely,

$$g(\varphi,\theta,\Psi) = \begin{bmatrix} \cos(\theta/2)\exp[\mathrm{i}(\varphi+\Psi)/2] & \mathrm{i}\sin(\theta/2)\exp[\mathrm{i}(\varphi+\Psi)/2] \\ \mathrm{i}\sin(\theta/2)\exp[\mathrm{i}(\Psi-\varphi)/2] & \cos(\theta/2)\exp[-\mathrm{i}(\varphi+\Psi)/2] \end{bmatrix} \tag{4.17}$$

The parameters $\mathbf{b}/n$ in $\omega_n(\mathbf{b}/n)$ are related to certain Euler angles, denoted β_1, β_2, and β_3, through the equation (both sides $\in$ SU(2))

$$\omega_n(\mathbf{b}/n) = g(\beta_1, \beta_2, \beta_3). \tag{4.18a}$$

The solution to this equation is

$$\cos(\beta_2) = \cos(b_1/n)\cos(b_2/n), \tag{4.18b}$$

$$\exp(\mathrm{i}\beta_1) = [\sin(\beta_2)]^{-1}[\sin(b_1/n)\cos(b_2/n) + \mathrm{i}\sin(b_2/n)], \tag{4.18c}$$

$$\exp[\mathrm{i}(\beta_1+\beta_3)/2] = \exp(\mathrm{i}b_3/2n)[\cos\beta_2/2)]^{-1} \times \\ \times [\cos(b_1/2n)\cos(b_2/2n) + \mathrm{i}\sin(b_1/2n)\sin(b_2/2n)] \tag{4.18d}$$

which follows from eqns V.3.2 (1)-V.3.3.(3). The point of this parameterization is that the hypergeometric functions associated with SU(2) take their most convenient form in terms of Euler angles.

The irreducible unitary representations T_l ($l = 0, \frac{1}{2}, 1, \ldots$), of SU(2) act on $(2l+1)$-dimensional Hilbert spaces $\mathbf{V}_l$ of $2l$-degree polynomials in one variable. The inner product is chosen so that

$$\{\Psi^l_m(s) = [(l-m)!(l+m)!]^{-\frac{1}{2}} s^{l-m} : |m| \leqslant l\} \tag{4.19a}$$

is an orthonormal basis for $\mathbf{V}_l$:

$$(\Psi^l_m, \Psi^l_j) = \delta_{mj}. \tag{4.19b}$$

Let $g \in$ SU(2); its matrix element with respect to this basis for $\mathbf{V}_l$ is written

$$t^l_{mj}(g) = (T_l(g)\,\Psi^l_j, \Psi^l_m). \tag{4.20a}$$

Vilenkin [1] gives a computation of the matrix elements for $g(\varphi, \theta, \Psi)$, e.g. V.3.3.1(4-7), V.3.3.3(6), and V.3.3.3(4). Abbreviating $t^l_{mj}[g(\beta_1, \beta_2, \beta_3)]$ by $t^l_{mj}(\beta_1, \beta_2, \beta_3)$, the cited equations lead to

$$\begin{aligned} t^l_{mm}(\beta_1, \beta_2, \beta_3) = {} & (2\pi)^{-1} \exp[-\mathrm{i}m(\beta_1 + \beta_3)] \times \\ & \times \int_0^{2\pi} [\cos(\beta_2/2) + \mathrm{i}\sin(\beta_2/2)\mathrm{e}^{-\mathrm{i}\gamma}]^{l-m} \times \\ & \times [\cos(\beta_2/2) + \mathrm{i}\sin(\beta_2/2)\mathrm{e}^{+\mathrm{i}\gamma}]^{l+m}\,\mathrm{d}\gamma; \end{aligned} \tag{4.20b}$$

we shall only need this special case, where $m = j$.

We shall also need the formula reducing tensor products into irreducible components. The basic formula is (V.3.8.1(6))

$$T_p(g) \otimes T_q(g) = \bigoplus_{l=|p-q|}^{p+q} T_l(g), \tag{4.21a}$$

from which one may deduce that

$$\bigotimes_1^n T_{\frac{1}{2}}(g) = \bigoplus_{j=0}^{n/2} M_{nj} T_j(g). \tag{4.21b}$$

The multiplicity function is

$$M_{nj} = n!(2j+1)![(n/2-j)!(n/2+j+1)!]^{-1}. \tag{4.21c}$$

The representation $\bigotimes_1^n T_{\frac{1}{2}}$, reduced in eqn (4.21b), is the pertinent representation for $\mathfrak{B}[n]$. It acts on $\mathscr{H}_n$, which also reduces:

$$\mathscr{H}_n = \bigoplus_{j=0}^{n/2} M_{nj} \mathbf{V}_j. \tag{4.22a}$$

Correspondingly, the trace reduces as well. In fact let A_j be an operator on $\mathbf{V}_j$ for $j = 0, \ldots, n/2$. Then if

$$A = \bigoplus_{j=0}^{n/2} M_{nj} A_j, \tag{4.22b}$$

it follows from the definition of orthogonal direct sums that

$$\operatorname{tr}(A) = \sum_{j=0}^{n/2} M_{nj} \operatorname{tr}_j(A_j). \tag{4.22c}$$

This result is applicable to the Hamiltonian which is 'diagonalized' by this decomposition. For the trace on $\mathbf{V}_j$ can be computed with respect to the orthonormal basis $\{\Psi^j_m : |m| \leqslant j\}$ defined in eqn (4.19a):

$$\operatorname{tr}_j A_j = \sum_{m=-j}^{+j} (\Psi^j_m, A_j \Psi^j_m), \tag{4.22d}$$

and we note that

$$T_j(\sigma^{(3)}) \Psi^j_m = m \Psi^j_m \tag{4.23a}$$

and

$$T_j(\sigma \cdot \sigma) \Psi^j_m = j(j+1) \Psi^j_m, \tag{4.23b}$$

the well-known quantum-mechanical angular momentum rules. Combining this with the definition of the Hamiltonian gives

$$(\Psi^j_{m'}, [H_n]_j \Psi^j_m) = E(n,j,m)\, \delta_{m'm}, \tag{4.24a}$$

where $[H_n]_j$ is the component of H_n which acts on $\mathbf{V}_j$ and E is the energy function

$$E(n,j,m) = (2m-n)\epsilon + (2g/n)\,[j(j+1) - m(m+1)]. \tag{4.24b}$$

Upon setting

$$\lambda(n,j,m) = \exp[-\beta E(n,j,m)], \tag{4.24c}$$

we can write down the generating function $F_n(\mathbf{b}/n)$ in terms of the hypergeometric function t^j_{mm} of eqn (4.20b),

$$\begin{aligned} F_n(\mathbf{b}/n) &= \phi^{(n)}_\beta [\omega_n(\mathbf{b}/n)] \\ &= \sum_{j=0}^{n/2} M_{nj} \sum_{m=-j}^{+j} \lambda(n,j,m)\, t^j_{mm}(\beta_1, \beta_2, \beta_3) \times \\ &\quad \times \left\{ \sum_{j=0}^{n/2} M_{nj} \sum_{m=-j}^{+j} \lambda(n,j,m) \right\}^{-1}. \end{aligned} \tag{4.25}$$

4.5. The global Gibbs state

Our task in this section is to show that the $\phi^{(n)}_\beta$ converge in a suitable sense to a state ϕ_β on the quasilocal algebra. This limit state is the global Gibbs state for the model.

We shall start by assuring ourselves that the pointwise limit of the generating functions $F_n(\mathbf{b})$ determines a state on $\mathfrak{B}$. In order for this to be so, all the parametric derivatives of the $F_n(\mathbf{b})$ must converge to the parametric derivatives of the limit function. We shall prove this by applying a theorem from the theory of complex functions.

THEOREM 4.1

If a sequence $\{f_n\}$ *of functions holomorphic on an open subset U of* $\mathbb{C}^p$ *converges pointwise on every compact subset of U to a limit function holomorphic on U, and if the family of functions is uniformly bounded on compacta, i.e. for every compact subset* $K \subset U$, $|f_n(\mathbf{z})| \leqslant M$ *(every* $\mathbf{z} \in K$*) independent of n, then all the derivatives converge pointwise on U:*

$$\lim_{n\to\infty} \frac{\partial^i}{\partial z_1^i} \cdots \frac{\partial^j}{\partial z_p^j} f_n(\mathbf{z}) = \frac{\partial^i}{\partial z_1^i} \cdots \frac{\partial^j}{\partial z_p^j} \lim_{n\to\infty} f_n(\mathbf{z}). \tag{4.26}$$

This theorem is the confluence of Vitali's theorem and Weierstrass' theorem (Narasimhan [1], Propositions 5 and 7).

Let us note that $t^j_{mm}(\beta_1, \beta_2, \beta_3)$ is holomorphic for all compact subsets of $\mathbb{C}^3$. For the Euler angles are holomorphic functions of the original parameters $\mathbf{b}/n$ on such sets; and the integral representation (4.20b) shows that t^l_{mm} is holomorphic on $\mathbb{C}^3$ compacta in the Euler angles. From eqn (4.25) we see that $F_n(\mathbf{b})$ is a sum of $(2j+1)\,n/2$ holomorphic functions, and so it, too, is holomorphic.

We next consider a uniform bound for the F_n. For $\mathbf{b}$ in $\mathbb{C}^3$, $\omega_n(\mathbf{b}/n)$ has the norm

$$\|\omega_n(\mathbf{b}/n)\| \leqslant \exp\{|\mathrm{Im}(b_1)| + |\mathrm{Im}(b_2)| + |\mathrm{Im}(b_3)|\}.$$

Let $\mathbf{b} \in K$, where K is the polydisc

$$K = \{|z_i| \leqslant \tfrac{1}{3}\ln(M): i = 1, 2, 3; 0 < \ln(M) < \infty\};$$

then in this region we have the bound

$$\|\omega_n(\mathbf{b}/n)\| \leqslant M.$$

Using the inequality

$$|\mathrm{tr}(\sigma A)| \leqslant |\mathrm{tr}(\sigma)| \,.\, \|A\|$$

valid for trace-class operators σ and bounded operators A on any separable Hilbert space, we set $\sigma = \sigma^{(n)}_\beta / \mathrm{tr}\,\sigma^{(n)}_\beta$ and $A = \omega_n(\mathbf{b}/n)$ to find

$$|F_n(\mathbf{b})| \leqslant M \qquad (\mathrm{b} \in \mathrm{K}).$$

Since it will turn out that $\lim_{n\to\infty} F_n(\mathbf{b}) = F(\mathbf{b})$ is holomorphic in compact subsets of $\mathbb{C}^3$, the above theorem applies. Thus $F(\mathbf{b})$ is the generating function for a state $\phi_\beta \in \mathfrak{S}(\mathfrak{B})$, with e.g.

$$\left(\frac{\partial^2}{\partial b_1 \partial b_2} F\right)(0) = \phi_\beta(J^{(1)}_p J^{(2)}_q). \tag{4.27}$$

Note that the mode indices (p and q) are arbitrary provided only that we do not fall foul of a commutation relation such as $J^{(+)}_p J^{(+)}_p = 0$. This reflects the permutation symmetry of the $\phi^{(n)}_\beta$; we shall check this symmetry for breaking after computing F.

Let us first find the limit of $t^l_{mm}(\beta_1, \beta_2, \beta_3)$ with large n. Actually we must take into account the fact that there is a sum over representations in F_n; this implies that the *n-limit* must be taken in such a way that

$$2l/n = y, 2m/n = w \tag{4.28a}$$

are constant and take values in the triangular range

$$\triangle = \{\mathbf{d} \in \mathbb{R}^2 : 0 \leqslant d_1 < 1; |d_2| \leqslant d_1\}. \tag{4.28b}$$

Granting this, we proceed as follows. First we find the leading terms in the Euler angles from eqns (4.18b)-(4.18d):

$$\begin{aligned} \beta_2 &\approx (b_1^2 + b_2^2)^{\frac{1}{2}}/n; \\ \exp[\mathrm{i}(\beta_1 + \beta_3)] &\approx \exp(\mathrm{i}b_3/n). \end{aligned} \tag{4.18e}$$

Then the limit of the hypergeometric function is $\lim_{l\to\infty} t^l_{mm}(\beta_1, \beta_2, \beta_3)$

$$= \exp(-\mathrm{i}wb_3) \int_0^{2\pi} \exp[\mathrm{i}(b_1^2 + b_2^2)^{\frac{1}{2}}(y\cos\gamma + \mathrm{i}w\sin\gamma)]\,\mathrm{d}\gamma/2\pi$$

$$= \exp(-\mathrm{i}wb_3) J_0\,[(y^2 - w^2)^{\frac{1}{2}}(b_1^2 + b_2^2)^{\frac{1}{2}}]; \tag{4.29}$$

J_0 is the Bessel function of the first kind, zeroth index. This result shows that this limit is related to the group contraction SU(2) to M(2), the Euclidean motions on the plane (cf. Inönü and Wigner (1), and Wilenkin [1], V.4.7, especially eqn (1)).

At this point, though, we must find the limit of the generating function F_n. Thirring (1) has done this, and the following are his results. Rewrite $F_n(\mathbf{b})$ in terms of y and w; this will give

$$F_n(\mathbf{b}) = \int_\triangle t^l_{mm}(\mathbf{b})\,\mathrm{d}\mu_n(y, w), \tag{4.30}$$

where the measure μ_n has its support in $\triangle$. Now as t^l_{mm} converges uniformly in compacta in $\mathbb{C}^3$ to a Bessel function (which is holomorphic), it is sufficient to consider the limit of the measures μ_n. The critical inverse temperature

$$\beta_c = \epsilon^{-1}\,\mathrm{arctanh}\,(\epsilon/g) \tag{4.31}$$

for this model is determined by this limit. Thirring (1) and Jelinek (1) find that

$$\lim_{n\to\infty} \mathrm{d}\mu_n(y, w) = \begin{cases} \delta(y - y_+)\,\delta(w - w_+)\,\mathrm{d}y\mathrm{d}w & \beta < \beta_c \\ \delta(y - y_-)\,\delta(w - w_-)\,\mathrm{d}y\mathrm{d}w & \beta \geqslant \beta_c, \end{cases} \tag{4.32a}$$

where the critical points in $\triangle$ are

$$y_+ = w_+ = \tanh(\beta\epsilon), \tag{4.32b}$$

and

$$y_- = (\beta g)^{-1}\mathrm{arctanh}(y_-); \quad w_- = \epsilon/g. \tag{4.32c}$$

The equation determining y_- is a transcendental equation typical of the BCS model and is known as the *gap equation*—more precisely, a variant of it is so known (Bardeen, Cooper, and Schrieffer (1)).

Keeping eqn (4.27) in mind, the generating function for the Gibbs state can be written down explicitly; it is

$$F(\mathbf{b}) = \begin{cases} \exp[ib_3 \tanh(\beta\epsilon)] & \beta<\beta_c \\ \exp(i\epsilon b_3/g) J_0[(y_-^2 - \epsilon^2/g^2)^{\frac{1}{2}}(b_1^2+b_1^2)^{\frac{1}{2}}] & \beta\geqslant\beta_c. \end{cases} \quad (4.33)$$

By differentiating F we can determine the thermal expectation values of the spin operators. It is convenient to treat the normal $[N]$ and superconducting $[S]$ states separately.

For the normal state it is useful to write $\mathbf{n} = (0, 0, 1) \in \mathbb{R}^3$ for the unit vector in the third direction, because we then have the concise formula

$$\phi_\beta[J_{p_1}^{(\alpha_1)} \ldots J_{p_r}^{(\alpha_r)}] = [\tanh(\beta\epsilon)]^r n_{\alpha_1} \ldots n_{\alpha_r} \quad (\beta<\beta_c), \quad (4.34)$$

which follows from differentiating $F(\mathbf{b})$ and may easily be verified.

From the obvious permutation symmetry of F we could have anticipated that in (4.34) the mode numbers (sites) $p_1, \ldots, p_r$ would not appear in the result.

For later use, when we come to the GNS representation, we shall need to know that the normalized solution vector in $\mathbb{C}^2$ to the eigenvalue equation

$$(\mathbf{n}\cdot\sigma)\xi = \xi \quad (4.35a)$$

is

$$\xi = (1, 0), \quad (4.35b)$$

which we write as $\hat{\mathrm{n}}_+$.

Now consider the superconducting phase. Using the integral representation of the Bessel function, for $\beta \geqslant \beta_c$ we may write (cf. eqn (4.29))

$$F(\mathbf{b}) = \int_0^{2\pi} d\gamma/2\pi \exp\{i[y_-^2 - w_-^2)^{\frac{1}{2}}(b_1\cos\gamma + b_2\sin\gamma) + b_3 w_-]\} \quad (\beta\geqslant\beta_c) \quad (4.36a)$$

Just as it was useful to introduce the vector $\mathbf{n} = (0, 0, 1) \in \mathbb{R}^3$ in order to write down the normal thermal averages of the spins, eqn (4.34), we also introduce the vector†

$$\mathbf{n}_{\beta\gamma} = ([1-w_-^2/y_-^2]^{\frac{1}{2}} \cos\gamma, \quad [1-w_-^2/y_-^2]^{\frac{1}{2}} \sin\gamma, \quad w_-/y_-) \quad (4.37)$$

in $\mathbb{R}^3$ to describe the superconducting thermal averages. For upon substituting $\mathbf{n}_{\beta\gamma}$ into eqn (4.36a) we find

$$F(\mathbf{b}) = \int_0^{2\pi} d\gamma/2\pi \exp[iy_- \mathbf{b}\cdot\mathbf{n}_{\beta\gamma})] \quad (\beta\geqslant\beta_c). \quad (4.36b)$$

† We shall write $\mathbf{n}_{\beta\gamma}$ or $\mathbf{n}(\beta,\gamma)$ for this vector, whichever is most convenient typographically.

Differentiating this now gives an expression for the thermal averages:

$$\phi_\beta[J_{p_1}^{(\alpha_1)} \ldots J_{p_r}^{(\alpha_r)}] = \int_0^{2\pi} \mathrm{d}\gamma/2\pi\,(y_-)^r n_{\alpha_1}(\beta,\gamma) \ldots n_{\alpha_r}(\beta,\gamma). \quad (\beta \geqslant \beta_c) \tag{4.38}$$

Note the angular integral; it indicates a spontaneous breakdown of gauge symmetry. We can also see that the permutation symmetry is preserved in the limit

$$\phi_\beta(\mathscr{S}A) = \phi_\beta(A) \tag{4.39}$$

for every β and every $A \in \mathfrak{B}$ (cf. Emch and Guenin (1)).

4.6. The Haag-Bogoliubov Hamiltonian

An important observation concerning ϕ_β in the superconducting state ($\beta \geqslant \beta_c$) is the angular integration as seen in eqn (4.38). We pose the following question, suggested by analogy with the Bose-Einstein condensation: Are there states $\{\psi_{\beta\gamma}\}$ on the quasilocal algebra $\mathfrak{B}$ whose angular composition is ϕ_β? In other words, the $\psi_{\beta\gamma}$ must satisfy

$$\phi_\beta = \int_0^{2\pi\,\oplus} \mathrm{d}\gamma/2\pi\;\psi_{\beta\gamma}. \tag{4.40}$$

These states $\psi_{\beta\gamma}$ would have generating functions F_γ such that

$$F(\mathbf{b}) = \int_0^{2\pi} \mathrm{d}\gamma/2\pi\; F_\gamma(\mathbf{b}); \quad (\mathbf{b} \in \mathbf{R}^3); \tag{4.41a}$$

eqn (4.36b) gives the affirmative result

$$F_\gamma(\mathbf{b}) = \exp[\mathrm{i}y_-\,\mathbf{b}\cdot\mathbf{n}_{\beta\gamma}]. \tag{4.41b}$$

A more difficult question is to relate this to local states $\psi_{\beta\gamma}^{(n)} \in \mathfrak{S}(\mathfrak{B}[\mathrm{n}])$ which are related somehow to the local Gibbs states. An important such set of states are the normal states whose density matrices are generated by the gauge-dependent Haag-Bogoliubov Hamiltonians:

$$H_{n\gamma} = -(2/\beta)\,\mathrm{arctanh}(y_-)\,\mathbf{L}_n\cdot\mathbf{n}_{\beta\gamma}. \tag{4.42}$$

An explicit computation of $\psi_{\beta\gamma}^{(n)}$ is possible, and up to order n^{-2} one finds

$$\begin{aligned}\psi_{\beta\gamma}^{(n)}[\omega_n(\mathbf{b}/n)] &= \mathrm{tr}_n[\exp(-\beta H_{n\gamma})\,\omega_n(\mathbf{b}/n)]/\mathrm{tr}_n[\exp(-\beta H_{n\gamma})] \\ &= \{1 + (\mathrm{i}y_-/n)\,\mathbf{b}\cdot\mathbf{n}_{\beta\gamma}\}^n + [O(n^{-2})]^n.\end{aligned} \tag{4.43}$$

Obviously

$$\lim_{n\to\infty} \psi_{\beta\gamma}^{(n)}[\omega_n(\mathbf{b}/n)] = F_\gamma(\mathbf{b}). \tag{4.44}$$

Historically, there were two reasons for considering $H_{n\gamma}$. The first was that $H_n - H_{n\gamma}$ is of order n^{-1}. It was felt that H_n ought to converge to the limit of $H_{n\gamma}$;

this is clearly untenable, and obscures the gauge symmetry breakdown. Another reason was that up to higher orders one has the implicit relation

$$H_{n\gamma} = \epsilon n - \epsilon L_n^{(3)} - (2g/n)\{\psi_{\beta\gamma}^{(n)}\left[L_n^{(+)}\right] L_n^{(-)} + \psi_{\beta\gamma}^{(n)}\left[L_n^{(-)}\right] L_n^{(+)}\}. \qquad (4.45)$$

In this form it is clear that this model is solvable because it is a *mean field model* (in the limit). That is, we replace certain of the fields $L_n^{(+)}$ by their means $\psi_{\beta\gamma}^{(n)}[L_n^{(+)}]$, so as to reduce the Hamiltonian to a linear form in the fields.

We mentioned above that the thermodynamic limit was related to the group contraction SU(2)→M(2), where M(2) is the group of Euclidean motions in the plane. That is, if $\mathbf{x} \in \mathbb{R}^2$ is a point in the plane, the generic M(2) element takes $\mathbf{x}$ to the point $\mathbf{x}_\alpha + \mathbf{c}_\alpha$ where $\mathbf{c} = (r\cos\delta, r\sin\delta) \in \mathbb{R}^2$ is a translation and $\mathbf{x}_\alpha$ is the vector obtained from $\mathbf{x}$ by a rotation through the angle α. The contractions leads to the convergence of the SU(2) matrix element $t^l_{mn}(\varphi, \theta, \Psi)$ to a matrix element of M(2):

$$\lim_{l\to\infty} t^l_{mm}(\delta, r/l, \alpha-\delta) = t^{\mathrm{i}\varphi}_{m-n,n}(r, \delta, \alpha). \qquad (4.46a)$$

The function $t^{\mathrm{i}\rho}_{pq}(r, \delta, \alpha)$ is the M(2) matrix element for the irreducible unitary representation $[\mathrm{i}\rho]$, where $\rho \in \mathbb{R}_* = \mathbb{R}\backslash\{0\}$; the parameters r, δ, α were described just above. Vilenkin [1] gives the result (V.4.3.1 (10′))

$$t^{i\rho}_{pq}(r, \delta, \alpha) = \mathrm{i}^{q-p}\exp\{-\mathrm{i}[q\alpha + (p-q)\delta]\} J_{q-p}(\rho r). \qquad (4.46b)$$

Of course, the sum over representations in the thermodynamic limit complicates matters for us. The point we wish to make here is that the corresponding Lie algebras contract and, as the Hamiltonian H_n is formed from $\mathbf{su}(2)$ elements, perhaps its contraction is interesting.

This algebraic contraction can be described by setting

$$\sigma_j^{(1)} = j\, d_j^{(1)},$$
$$\sigma_j^{(2)} = j\, d_j^{(2)},$$
$$\sigma_j^{(3)} = d_j^{(3)}, \qquad (4.47)$$

and taking the limit $j \to \infty$. The symbol $\sigma_j^{(\alpha)}$ stands for the representation of $\sigma^{(\alpha)}$ corresponding to the representation T_j of SU(2); the $d_j^{(\alpha)}$ have a similar meaning for M(2) with the $d^{(\alpha)}$ generating its Lie algebra. With this identification it follows that

$$L_n^{(1)} = \bigoplus_{j=0}^{n/2} jM_{nj}\, d_j^{(1)},$$
$$L_n^{(2)} = \bigoplus_{j=0}^{n/2} jM_{nj}\, d_j^{(2)},$$
$$L_n^{(3)} = \bigoplus_{j=0}^{n/2} M_{nj}\, d_j^{(3)}. \qquad (4.48)$$

If one substitutes this decomposition into H_n, there is now a non-trivial scale factor j to separate the part which survives contraction from the part which does not. When the intensive variable parameters y_+ and w_+ are accounted for, one may conjecture that it is $H_{n\gamma}$ which is the part of H_n which survives the (generalized) contraction. It would be interesting for this conjecture to be examined, for if it is true it might be immediately applicable to the various maser models (Davies (3); Hepp and Lieb (1,2); Lieb (1)). In all these cases there is the familiar complication that a sum over representations occurs, and so one has to perform a generalized contraction. And finally, in this regard one may conjecture that this phenomenon of group contraction is present because these are mean field models. For example, the quantity

$$M(\phi_\beta^{(n)}) = \phi_\beta^{(n)}(J_p^{(-)}[n]\, J_q^{(-)}[n]) + \phi_\beta^{(n)}(J_p^{(-)}[n])\, \phi_\beta^{(n)}(J_q^{(-)}[n]) \tag{4.49a}$$

may serve as a measure of the polarizability of the state $\phi_\beta^{(n)}$, and it is non-zero. But in the limit it approaches zero, showing clearly that there are fewer 'degrees of freedom' in the limit. In terms of $\psi_{\beta\gamma}^{(n)}$, the polarizability dispersion of eqn (4.49a) can be computed, and it is

$$M(\psi_{\beta\gamma}^{(n)}) = (y_-)^2\,[n_1(\beta,\gamma) + \mathrm{i}n_2(\beta,\gamma)]^2/n; \tag{4.49b}$$

this leads to the interesting result that $M(\psi_{\beta\gamma}^{(n)})$ approaches zero as n^{-1} in the thermodynamic limit.

4.7. The thermodynamic representation

In this section we shall exhibit the GNS representations associated with the states $\phi_\beta \sim [\mathscr{H}_\beta, \pi_\beta, \Omega_\beta]$, and for $\beta \geqslant \beta_c$ those associated with $\psi_{\beta\gamma} \sim [\mathscr{H}_{\beta\gamma}, \pi_{\beta\gamma}, \Omega_{\beta\gamma}]$. Let us start with the normal states ϕ_β (for $\beta < \beta_c$).

In eqn (4.35) we introduced vectors $\mathbf{n} = (0,0,1) \in \mathbb{R}^3$ and $\hat{\mathrm{n}}_+ = (0,1) \in \mathbb{C}^2$ which are related through $(\sigma\cdot\mathbf{n})\,\hat{\mathrm{n}}_+ = \hat{\mathrm{n}}_+$. We shall also need the vector $\hat{\mathrm{n}}_- = (0,1) \in \mathbb{C}^2$; this vector satisfies the eigenvalue equation $(\sigma\cdot\mathbf{n})\,\hat{\mathrm{n}}_- = -\hat{\mathrm{n}}_-$.

Now we construct the normalized fiducial vector $\xi_\beta \in \mathbb{C}^2 \otimes \mathbb{C}^2$,

$$\xi_\beta = [(1+y_+)/2]^{\frac{1}{2}}\,\hat{\mathrm{n}}_+ \otimes \hat{\mathrm{n}}_+ + [(1-y_+)/2]^{\frac{1}{2}}\,\hat{\mathrm{n}}_- \otimes \hat{\mathrm{n}}_-. \tag{4.50}$$

We use this vector to construct an infinite tensor product Hilbert space as follows.

Let $\mathscr{H}_\beta^{(n)}$ be the n-fold tensor product of $\mathbb{C}^2 \otimes \mathbb{C}^2$ with itself,

$$\mathscr{H}_\beta^{(n)} = \bigotimes_1^n (\mathbb{C}^2 \otimes \mathbb{C}^2), \tag{4.51a}$$

and identify $\mathscr{H}_\beta^{(n)}$ with a subspace of $\mathscr{H}_\beta^{(m)}$ for $n < m$ by means of the mapping

$$a_{nm} : w^{(n)} \mapsto w^{(n)} \bigotimes_{n+1}^{m} (\xi_\beta). \tag{4.51b}$$

The Hilbert space inductive limit of this family is the GNS representation space

$$\mathscr{H}_\beta = \varinjlim \{a_{nm}[\mathscr{H}_\beta^{(n)}] : n, m \in \mathbf{N};\, n < m\}. \tag{4.52a}$$

The distinguished unit product vector Ω_β is $(\beta < \beta_c)$

$$\Omega_\beta = \underset{\mathbb{N}}{\otimes} \xi_\beta; \tag{4.53}$$

one sometimes writes

$$\mathscr{H}_\beta = \underset{\mathbb{N}}{\otimes}^{\Omega_\beta}(\mathbb{C}^2 \otimes \mathbb{C}^2) \tag{4.52b}$$

in place of (4.52a).

Omitting subscripts on unit operators which show what spaces they act on, the representation

$$\pi_\beta : \mathfrak{B} \to \mathbb{B}(\mathscr{H}_\beta)$$

is (the algebraic morphism) determined by

$$\pi_\beta\left[J_p^{(\alpha)}\right] = \overset{p-1}{\underset{1}{\otimes}} \mathbb{1} \otimes (\sigma^{(\alpha)} \otimes \mathbb{1}) \overset{\infty}{\underset{p+1}{\otimes}} \mathbb{1}. \tag{4.54}$$

Let us do a simple computation which is typical of the way in which eqns (4.52)-(4.54) can be verified. First we note that

$$(\sigma^{(1)} \otimes \mathbb{1})\, \xi_\beta = [(1 - y_+)/2]^{\frac{1}{2}}\, \hat{\mathbf{n}}_- \otimes \hat{\mathbf{n}}_+ + [(1 - y_+)/2]^{\frac{1}{2}}\, \hat{\mathbf{n}}_+ \otimes \hat{\mathbf{n}}_-,$$

so that

$$(\xi_\beta, [\sigma^{(1)} \otimes \mathbb{1}]\, \xi_\beta) = 0.$$

Similarly,

$$(\xi_\beta, [\sigma^{(2)} \otimes \mathbb{1}]\, \xi_\beta) = 0,$$

but as

$$(\sigma^{(3)} \otimes \mathbb{1})\, \xi_\beta = [(1 + y_+)/2]^{\frac{1}{2}}\, \hat{\mathbf{n}}_+ \otimes \hat{\mathbf{n}}_+ - [(1 - y_+)/2]^{\frac{1}{2}}\, \hat{\mathbf{n}}_- \otimes \hat{\mathbf{n}}_-,$$

we find that

$$(\xi_\beta, [\sigma^{(3)} \otimes \mathbb{1}]\, \xi_\beta) = y_+.$$

These results can be subsummed under the formula

$$(\xi_\beta, [\sigma^{(\alpha)} \otimes \mathbb{1}]\, \xi_\beta) = y_+ n_\alpha, \tag{4.55}$$

since $n_\alpha = \delta_{\alpha 3}$.

Now let $p_1 < p_2 < \ldots < p_r$ be mode indices. Then

$$\pi_\beta\left[J_{p_1}^{(\alpha_1)} \ldots J_{p_r}^{(\alpha_r)}\right] =$$

$$\overset{p_1-1}{\underset{1}{\otimes}} \mathbb{1} \otimes [\sigma^{(\alpha_1)} \otimes \mathbb{1}] \overset{p_2-1}{\underset{p_1+1}{\otimes}} \mathbb{1} \otimes [\sigma^{(\alpha_2)} \otimes \mathbb{1}] \otimes$$

$$\otimes \ldots \otimes [\sigma^{(\alpha_r)} \otimes \mathbb{1}] \overset{\infty}{\underset{p_r+1}{\otimes}} \mathbb{1} \tag{4.56a}$$

follows directly from (4.54). Next we operate with this on Ω_β, and then take the inner product of the result with Ω_β:

$$\phi_\beta\left[J_{p_1}^{(\alpha_1)} \ldots J_{p_r}^{(\alpha_r)}\right]$$

$$= (\Omega_\beta, \pi_\beta \left[J_{p_1}^{(\alpha_1)} \ldots J_{p_r}^{(\alpha_r)}\right] \Omega_\beta)$$

$$= \prod_{i=1}^{r} (\xi_\beta [\sigma^{(\alpha_i)} \otimes \mathbb{1}] \xi_\beta)$$

$$= (y_+)^r n_{\alpha_1} \ldots n_{\alpha_r}, \tag{4.56b}$$

which is the known result (4.34). We leave the details of the analogous calculation when there are coincident mode numbers (sites) to the reader.

Even though there is an angular integration in the formula for the thermal averages in the superconducting state, the analysis for $\psi_{\beta\gamma}$ is analogous to the one above.

In eqn (4.37) we introduced the vector $\mathbf{n}(\beta, \gamma) \in \mathbb{R}^3$. The two normalized $\mathbb{C}^2$ eigenvectors

$$[\sigma \cdot \mathbf{n}(\beta, \gamma)]\, \hat{\mathbf{n}}_\pm(\beta, \gamma) = \pm\, \hat{\mathbf{n}}_\pm(\beta, \gamma) \tag{4.57a}$$

are

$$\hat{\mathbf{n}}_\pm(\beta, \gamma) = [(1 \mp L_-)/2]^{\frac{1}{2}} \left(\mathrm{e}^{-\mathrm{i}\gamma}[(1 \pm L_-)/(1 \mp L_-)]^{\frac{1}{2}}, 1\right) \tag{4.57b}$$

where $L_- = w_-/y_-$. The fiducial vector $\xi_{\beta\gamma}$ for this state is formally similar to ξ_β for the normal state (4.50), as indeed are the subsequent formulae. Then

$$\xi_{\beta\gamma} = [(1 + y_-)/2]^{\frac{1}{2}}\, \hat{\mathbf{n}}_+ (\beta, \gamma) \otimes \hat{\mathbf{n}}_+ (\beta, \gamma) + \\ + [(1 - y_-)/2]^{\frac{1}{2}}\, \hat{\mathbf{n}}_- (\beta, \gamma) \otimes \hat{\mathbf{n}}_- (\beta, \gamma). \tag{4.58a}$$

Defining the distinguished normalized product vector

$$\Omega_{\beta\gamma} = \bigotimes_{\mathbb{N}} (\xi_{\beta\gamma}), \tag{4.58b}$$

the representation Hilbert space is

$$\mathscr{H}_{\beta\gamma} = \bigotimes_{\mathbb{N}}^{\Omega_{\beta\gamma}} (\mathbb{C}^2 \otimes \mathbb{C}^2). \tag{4.59}$$

The representation is

$$\pi_{\beta\gamma} : \to \mathbb{B}(\mathscr{H}_{\beta\gamma})$$

$$\pi_{\beta\gamma}\, [J_p^{(\alpha)}] = \bigotimes_1^{p=1} \mathbb{1} \otimes (\sigma^{(\alpha)} \otimes \mathbb{1}) \bigotimes_{p+1}^{\infty} \mathbb{1}. \tag{4.60}$$

The nature of the constructions are such that the same computation which verified the representation $[\mathscr{H}_\beta, \pi_\beta, \Omega_\beta]$ for ϕ_β when $\beta < \beta_c$ will suffice to verify these $[\mathscr{H}_{\beta\gamma}, \pi_{\beta\gamma}, \Omega_{\beta\gamma}] \sim \psi_{\beta\gamma}$ when $\beta \geqslant \beta_c$.

The full Gibbs state representations follow by direct integration:

$$\Omega_\beta = \int_0^{2\pi \oplus} \mathrm{d}\gamma/2\pi \bigotimes_{\mathbb{N}} (\xi_{\beta\gamma}),$$

$$H_\beta = \int_0^{2\pi \oplus} \mathrm{d}\gamma/2\pi \bigotimes_{\mathbb{N}}^{\Omega_{\beta\gamma}} (\mathbb{C}^2 \otimes \mathbb{C}^2),$$

$$\pi_\beta (A) = \int_0^{2\pi \oplus} \mathrm{d}\gamma/2\pi\; \pi_{\beta\gamma}(A) \quad (A \in \mathfrak{B}). \tag{4.61}$$

Actually, the zero temperature case is singular, so these above formulae hold only for $\beta_c < \beta < \infty$. As y_-/w_- is well behaved in the limit, we can define $\hat{\mathfrak{n}}_+(\infty, \gamma)$. Using this, we construct the normalized product vector

$$\Omega_{\infty\gamma} = \bigotimes_{\mathbf{N}} \hat{\mathfrak{n}}_+(\infty, \gamma) \tag{4.62a}$$

and use this to construct the Hilbert space

$$\mathscr{H}_{\infty\gamma} = \bigotimes_{\mathbf{N}}^{\Omega_{\infty\gamma}} (\mathbb{C}^2). \tag{4.63a}$$

The representation is given by

$$\pi_{\infty\gamma}[J_p^{(\alpha)}] = \bigotimes_{1}^{p-1} \mathbb{1} \otimes \sigma^{(\alpha)} \bigotimes_{p+1}^{\infty} \mathbb{1}. \tag{4.64a}$$

The Gibbs state at zero temperature is again a direct integral:

$$\Omega_\infty = \int_0^{2\pi\,\oplus} \frac{d\gamma}{2\pi} \bigotimes_{\mathbf{N}} \hat{\mathfrak{n}}_+(\infty, \gamma), \tag{4.62b}$$

$$\mathscr{H}_\infty = \int_0^{2\pi\,\oplus} \frac{d\gamma}{2\pi} \bigotimes_{\mathbf{N}}^{\Omega_{\infty\gamma}} (\mathbb{C}^2), \tag{4.63b}$$

and

$$\pi_\infty = \int_0^{2\pi\,\oplus} \frac{d\gamma}{2\pi} \pi_{\infty\gamma}(A) \qquad (A \in \mathfrak{B}). \tag{4.63c}$$

From standard results on the factor types of tensor products of C^*-algebras (Sakai [1], §4.4, especially proposition 4.4.7), Jelinek (1) has shown that the types associated with the global Gibbs states are:

$\pi_\beta(\mathfrak{B})''$ is (a) a Type-II_1 factor for $\beta = 0$;
(b) a Type-III factor for $0 < \beta < \beta_c$;
(c) a direct integral of Type-III factors for $\beta_c \leqslant \beta \leqslant \infty$;
(d) a direct integral of irreducible representations for $\beta = \infty$.

4.8. Time translations

The local time translations formed from the BCS Hamiltonian form automorphism groups $\tau^{(n)}(\mathbf{R})$ of the $\mathfrak{B}[n]$. But there is no hope of the limit $\lim_{n\to\infty} \tau^{(n)}$ forming an automorphism group of $\mathfrak{B}$. To see this it suffices to consider e.g.

$$[J_p^{(3)}[n], H_n]_- = -4g/n\,(J_p^{(+)}[n]\,L_n^{(-)} - L_n^{(+)}\,J_p^{(-)}[n]) \qquad (p < n). \tag{4.64}$$

The trouble is that the operators $L_n^{(\alpha)}/n$ do not converge uniformly as $n \to \infty$. As it follows from this commutator that $L_n^{(\alpha)}/n$ will appear in $\tau_t^{(n)}\{J_p^{(3)}[n]\}$, the limit cannot be an element of $\mathfrak{B}$. However, as $\phi_\beta(L_n/n) \to y_+ n_-$ for $\beta < \beta_c$ and $\psi_{\beta\gamma}(L_n/n) \to y_- n_{\beta\gamma}$ for $\beta \geqslant \beta_c$, we can expect relatively decent behaviour in the thermodynamic representations, and this is what does occur.

For the normal region, DS I and II hold for ϕ_β; for the superconducting region they hold for $\psi_{\beta\gamma}$. Direct integration of the latter limits show that they hold for $\phi_\beta, \beta \geqslant \beta_c$. We shall not prove this as such; the proofs are implicit in Thirring (1) and Jelinek (1). What we shall do is examine the time-dependence in the thermodynamic representations.

Let us start with the more interesting region $\beta \geqslant \beta_c$. The Haag-Bogoliubov Hamiltonian $H_{n\gamma}$ generates a unitary group $U_t^{(n\gamma)} = \exp(\mathrm{i}t\, H_{n\gamma})$ which implements an automorphism group on $\mathfrak{B}[n]$:

$$\begin{aligned} \tau^{(n\gamma)} &: \mathbb{R} \rightarrow \mathrm{Aut}(\mathfrak{B}[n]); \\ \tau_t^{(n\gamma)}[A] &= U_t^{(n\gamma)} A\, U_{-t}^{(n\gamma)}. \end{aligned} \tag{4.65}$$

Things will be easier if we work with the operator $\exp(\mathbf{b}\cdot\mathbb{L}_n/n)$ rather than $\omega_n(\mathbf{b}/n)$. Using the known properties of angular momentum operators, one may verify that

$$\begin{aligned} &\tau_t^{(n\gamma)}[\exp(\mathbf{b}\cdot\mathbb{L}_n/n)] \\ &= \exp\{[b + 2t\beta^{-1}\operatorname{arctanh}(y_-)\, \mathbf{n}_{\beta\gamma} \times \mathbf{b}]\,.\mathbb{L}_n/n\}, \end{aligned} \tag{4.66}$$

which we write as $\exp(\mathbf{b}_t\cdot\mathbb{L}_n/n)$. Using eqn (4.60) which gives $\pi_{\beta\gamma}[J_p^{(\alpha)}]$ explicitly, we find

$$\begin{aligned} &\pi_{\beta\gamma}\{\tau_t^{(n\gamma)}[\exp(\mathbf{b}\cdot\mathbb{L}_n/n)]\} \\ &= \bigotimes_{p=1}^{n} \exp(\mathbf{b}_t\cdot\mathbf{J}_p/n) \bigotimes_{n+1}^{\infty} \mathbb{1}. \end{aligned} \tag{4.67}$$

This formula is explicit enough so that one may find the limit of the $\tau^{(n\gamma)}$:

$$(\text{str. } \mathscr{H}_{\beta\gamma}) - \lim_{n\to\infty} \pi_{\beta\gamma}[\tau_t^{(n\gamma)}(A)] = \tau_t^{(\beta\gamma)}\pi_{\beta\gamma}(A) \quad (A \in \mathfrak{B}_{\mathrm{L}})$$

with

$$\tau^{(\beta\gamma)} : \mathbb{R} \rightarrow \mathrm{Aut}(\mathfrak{B}_{\beta\gamma}'').$$

Similarly implemented on $\mathscr{H}_{\beta\gamma}$ by the unitary group (cf. Jelinek (1), eqn (8)),

$$\begin{aligned} U_t^{(\beta\gamma)} = \bigotimes_{\mathbb{N}} [&\exp\{\mathrm{i}tgy_-(1 - \mathbf{J}\cdot\mathbf{n}_{\beta\gamma})\} \otimes \\ &\otimes \exp\{-\mathrm{i}tgy_-(1 - \mathbf{J}\cdot\mathbf{n}_{\beta\gamma})\}]. \end{aligned} \tag{4.68a}$$

For the superconducting Gibbs state, DS I and II hold in a representation-dependent way:

$$\begin{aligned} &\lim_{n\to\infty} \phi_\beta^{(n)}[\tau_{t_1}^{(n)}(A_1) \ldots \tau_{t_j}^{(n)}(A_j)] \\ &= \int_0^{2\pi} \mathrm{d}\gamma/2\pi\, \Psi_{\beta\gamma}\{\tau_{t_1}^{(\beta\gamma)}[\pi_{\beta\gamma}(A_1)] \ldots \tau_{t_j}^{(\beta\gamma)}[\pi_{\beta\gamma}(A_j)]\}, \end{aligned}$$

$$\begin{aligned} &\lim_{m\to\infty} \lim_{n\to\infty} \phi_\beta^{(n)}[\tau_{t_1}^{(n)}(A_1) \ldots \tau_{t_j}^{(n)}(A_j)\, \tau_{s_1}^{(m)}(B_1) \ldots \tau_{s_l}^{(m)}(B_l)] \\ &= \int_0^{2\pi} \mathrm{d}\gamma/2\pi\, \Psi_{\beta\gamma}\{\tau_{t_1}^{(\beta\gamma)}[\pi_{\beta\gamma}(A_1)] \ldots \tau_{t_j}^{(\beta\gamma)}[\pi_{\beta\gamma}(A_j)]\, \tau_{s_1}^{(\beta\gamma)}[\pi_{\beta\gamma}(B_1) \ldots \end{aligned}$$

$$\ldots \tau_{s_l}^{(\beta\gamma)}[\pi_{\beta\gamma}(B_l)]\}, \tag{4.69}$$

where $A_1, \ldots, A_j \in \mathfrak{B}[n]$, $B_1, \ldots, B_l \in \mathfrak{B}[m]$ and $\Psi_{\beta\gamma}$ is the unique continuous extension of $\psi_{\beta\gamma}$ to a state on $\mathfrak{B}''_{\beta\gamma}$:

$$\Psi_{\beta\gamma}(Q) = (\Omega_{\beta\gamma}, Q\Omega_{\beta\gamma}) \qquad (Q \in \mathfrak{B}''_{\beta\gamma}). \tag{4.70}$$

It then follows from the full DS theory that (Dubin and Sewell (1))

$$\tau_t^{(\beta)} = \int_0^{2\pi \oplus} \frac{d\gamma}{2\pi}\, \tau_t^{(\beta\gamma)} \tag{4.71}$$

defines an automorphism group

$$\tau^{(\beta)} : \mathbb{R} \to \mathrm{Aut}(\mathfrak{B}''_\beta)$$

which is unitarily implemented by the unitary group

$$U_t^{(\beta)} = \int_0^{2\pi \oplus} \frac{d\gamma}{2\pi}\, U_t^{(\beta\gamma)}. \tag{4.68b}$$

At this point we can clear up in just what way $\lim_{n\to\infty} H_n$ and $\lim_{n\to\infty} H_{n\gamma}$ coincide, namely,

$$(\text{str. } \mathscr{H}_{\beta\gamma})\text{-}\lim_{n\to\infty} \pi_{\beta\gamma}[\tau_t^{(n)}(A)] = \tau_t^{(\beta\gamma)}[\pi_{\beta\gamma}(A)] \tag{4.72}$$

for every $A \in \mathfrak{B}$. So we see that the limits coincide on $\mathfrak{B}''_{\beta\gamma}$. Since $H_{n\gamma}$ is γ-independent, whereas H_n is not, these results explain the difficulties people found in attempting to compare the limits on the quasilocal algebra $\mathfrak{B}$ rather than on $\mathfrak{B}''_{\beta\gamma}$ (Thirring and Wehrl (1)).

For the normal region, DS I and II hold without any angular integration. The time-translation automorphism group

$$\tau_\beta : \mathbb{R} \to \mathrm{Aut}(\mathfrak{B}''_\beta) \quad (\beta < \beta_c)$$

is unitarily implemented by the unitary group (cf. eqn (4.68)

$$U_t^{(\beta)} = \bigotimes_{\mathbf{N}} \{\exp[\mathrm{i}tgy_+(1 - \mathbf{J}_p\cdot\mathbf{n})] \otimes \exp[-\mathrm{i}tgy_+(1 - \mathbf{J}_p\cdot\mathbf{n})]\}. \tag{4.73}$$

For the normal region, $0 < \beta < \beta_c$, $\mathfrak{B}''_\beta$ is a Type-III factor, and $\mathfrak{B}''_0$ is a Type-II$_1$ factor. From the general Tomita-Takesaki theory of modular Hilbert algebras (Takesaki [1]), one knows that ϕ_β is the unique extremal KMS state.

Now the $\psi_{\beta\gamma}^{(n)}$ (resp. $\phi_\beta^{(n)}$) are $(\beta, \tau^{(n\gamma)})$-KMS (resp. $(\beta, \tau^{(n)})$-KMS). The quoted convergence results imply that the limiting states $\psi_{\beta\gamma}$ (resp. ϕ_β) are $(\beta, \tau^{(\beta\gamma)})$-KMS (resp. $(\beta, \tau^{(\beta)})$-KMS). Moreover, as the $\mathfrak{B}''_{\beta\gamma}$ are Type-III factors for $\beta_c < \beta < \infty$, in this region they must be the extremal KMS states, and the integral decomposition

$$\Phi_\beta = \int_0^{2\pi \oplus} \frac{d\gamma}{2\pi}\, \Psi_{\beta\gamma} \tag{4.74}$$

must be both the extremal KMS and the central decomposition simultaneously.

It remains to give a meaning to the angular decomposition. We may take $L_n^{(3)}$ as the dynamic variable corresponding to the number of Cooper pairs. The automorphism group relating to this is (recall that T is the additive group of angles)

$$\Gamma_n : T \to \mathrm{Aut}(\mathfrak{B}[n]),$$
$$\Gamma_n(\theta)[A] = \exp(\mathrm{i}\theta L_n^{(3)})A \exp(-\mathrm{i}\theta L_n^{(3)}). \tag{4.75}$$

Some calculations using σ-matrices lead to

$$\Gamma_n(\theta)\,[H_n] = H_n, \tag{4.76a}$$
$$\Gamma_n(\theta)\,[H_{n\gamma}] = -[2\,\mathrm{arctanh}(y_-)/\beta]\,\mathbf{n}'_{\beta\gamma}\cdot\mathbf{L}_n, \tag{4.76b}$$

and

$$\mathbf{n}'_{\beta\gamma} = \mathbf{n}_{\beta\gamma} + (2\theta)\,\mathbf{n}_{\beta\gamma} \times \mathbf{n}, \tag{4.76c}$$

where $\mathbf{n} = (0, 0, 1) \in \mathbf{R}^3$ as usual.

It is clear that if the automorphism converges in the representations—and it does—then Φ_β will be $\Gamma^{(\beta)}(T)$-invariant but $\Psi_{\beta\gamma}$ will not be $\Gamma^{(\beta\gamma)}(T)$-invariant. Thus it is this $\Gamma(T)$ gauge symmetry which is spontaneously broken, and signals the presence of a superconducting component. Just as for the superfluid phase in the Bose gas it is the number operator which misbehaves. One finds the statement that the vacuum for these models is continuously degenerate; this is another aspect of these hyperussiastic states being 'macroscopic'. Finally let us note that if L_n^3/n is considered as a density operation for Cooper pairs, in the normal region the Cooper pair density is $\tanh(\beta\epsilon)$ and in the superconducting region it is ϵ/g. Remarkably, this latter density is temperature-independent.

5

Lattice gases

5.1. Introduction

In the previous chapter we described the Cooper electron pairs by means of spin operators. The fact that electrons possess an intrinsic spin was proposed by Uhlenbeck and Goudsmit as early as 1925 in order to account for certain features of atomic spectra (line splitting) (Born [1]; Uhlenbeck and Goudsmit (1,2); Pauli (2,3)). As such line splittings are affected by magnetic fields, there must be a spin-magnetic-field interaction. There is also a smaller spin-spin interaction, having no classical analogy; this so-called 'exchange force' is an important constituent of chemical binding (Heisenberg (1,2)).

These effects depend on the intrinsic electron spin (Dirac (1,2); Uhlenbeck and Goudsmit (1,2)). It was Heisenberg (2) who pointed out that clusters of atoms of certain metallic elements behaved collectively so as to give rise to an effective spin variable. The age-old mystery of the origins of magnetism was seen to be a quantum-mechanical effect. For these clusters of atoms, known as magnetic domains, can account for many features of magnetism in metals. Although a full account of the magnetic properties of matter requires the solution of a complicated many-body problem, certain idealized models are known which have the correct qualitative behaviour, at least to the extent that their study is worthwhile (Mattis [1]).

All of these models have the following similar features. The magnetic domains of the real material are mathematically described by a discrete countable set (lattice) of Hilbert spaces, one for each lattice site. The vectors of these Hilbert spaces correspond to states of quantized spin vibration, and so the spaces are often called 'spin spaces'.

Although one can construct a mathematical model corresponding to any half-integral spin $s \in \frac{1}{2}\mathbf{N}$, it is the case $s = \frac{1}{2}$ which is most important. For spin s, each lattice-site Hilbert space is a copy of (isomorphic to) the $(2s + 1)$-dimensional space $\mathbf{C}^{2s+1}$; therefore we shall mostly consider structures associated with the space $\mathbf{C}^2$. As regards physical interpretation, the vectors $(1, 0)$ and $(0, 1)$ of $\mathbf{C}^2$ are taken to correspond to a spin aligned up and down respectively.

The spin operators are the Pauli matrices in this formalism: the matrices $\sigma^{(\pm)} = \sigma^{(1)} \pm \mathrm{i}\, \sigma^{(2)}$ are then spin-flip operators; $\sigma^{(+)}$ flipping a spin from down to up, for example.

Heisenberg (2) proposed a certain quadratic form in the spin operators as a model magnetic Hamiltonian. This Heisenberg Hamiltonian can be derived from a certain approximation to the effective spin part of the full many-body Hamiltonain (Mattis [1]; Van Vleck [1]), but we shall take the Heisenberg and other spin Hamiltonians as given inputs here.

There have been some deep results proven for the Heisenberg model (cf. e.g. Mattis [1]; Robinson (3); Streater (1)), but an exact solution seems beyond reach by at least an order of magnitude. In an attempt to find exact results, various simpler approximations to the Heisenberg Hamiltonian have been examined. The Ising model (Brush [1]) follows from the Heisenberg model by considering only $\sigma^{(3)}$ terms in both the Hamiltonian and the operator algebra of spins; the generalized Ising model keeps the full $\mathfrak{su}(2)$ spin algebra, but only $\sigma^{(3)}$ terms in the Hamiltonian. The *XY*-model, on the other hand, retains only the $\sigma^{(1)}$ and $\sigma^{(2)}$ terms from the full Heisenberg model.

Even these truncated models are not exactly solvable in fullest generality; one must restrict the number of lattice dimensions and the range of the interactions. In this chapter our main concern will be to examine the algebraic aspects of the two-dimensional Ising model, first solved by Onsager (1). At all crucial computational points we shall refer to the literature for details and simply quote the results, for the calculations are formidably long and to a large extent are known from the non-algebraic analysis of the model (Huang [1]; Mattis [1]; Schultz, Mattis, and Lieb (1)). There are some new computations, particular to the algebraic analysis (Marinaro and Sewell (1)); the details of these are analogous to the non-algebraic ones, and are also left to the references.

As the reader will see, our treatment of the Ising model ends with the demonstration of the spontaneous breakdown of spin-reversal symmetry for the global Gibbs state. The two-dimensional Ising model is known to have a non-zero spontaneous magnetization and a logrithmic singularity in its specific heat capacity. As these are the most important physical aspects of the model, we must justify their omission here.

The usual computation of these thermodynamic quantiti^s proceeds by calculating their finite volume counterparts, usually through the $\mathfrak{B}[n]$ grand partition function $E_n^{(\beta)}(1_n)$ (cf. eqn (5.43) below). Having found the free energy density, say, the limit $n \to \infty$ is taken, and the quasilocal free energy density corresponding to the limiting regions is found. The limit function could depend upon the manner in which the limit is taken; on boundary conditions, for instance. But in any event, such a computation is not *algebraic*, and the methods described in this book do not add anything to it.

If one knew the global Gibbs state explicitly, one could use that knowledge to derive independently such thermodynamic quantities. As the Gibbs state is locally normal, its restriction to any local subalgebra, say $\phi_{\beta|n}$ on $\mathfrak{B}[n]$, is given through a density matrix:

$$\phi_{\beta|n}(A) = \mathrm{tr}(\sigma_{\beta n}A) \quad (\forall A \in \mathfrak{B}[n]).$$

There is a vexing question of precisely how this restriction ought to be taken; in particular, what boundary conditions to choose for the generator of $\sigma_{\beta n} = \exp(-\beta D_{\beta n})$. For $D_{\beta n}$ generally will not be the same operator as the local Hamiltonian for $\mathfrak{B}[n]$. After the choice of $D_{\beta n}$ has been made, the quantities

$$S_{\beta n} = -(k/n)\,\mathrm{tr}[\sigma_{\beta n} \ln(\sigma_{\beta n})]$$

and

$$u_{\beta n} = (1/n)\,\mathrm{tr}[\sigma_{\beta n} D_{\beta n}]$$

are the entropy density and internal energy density corresponding to ϕ_β restricted to the first n modes of the lattice. In the same way the restriction to arbitrary modes can be derived. Similarly, one can also analyse the Gibbs state for the other models in this book. We do not do this for the Ising model because ϕ_β is not known explicitly enough. We have not done this for the previous three models because we do not know enough about the restriction process.

The Ising model is still the object of research and there is an extensive modern literature on the subject. Two recent papers (Abraham, Gallavotti, and Martin-Löf (1), and Martin-Löf (1)) will serve the reader as a starting point for further material on the Ising model. There are also many papers on the general theory of quantum-lattice gases from an algebraic point of view. Some references are Araki (5), Brascamp (1), Dubin (1), Dubin and Streater (1), Ginibre, Grossmann, and Ruelle (1), and Lanford and Robinson (1, 2).

5.2. Spin-lattice kinematics

A lattice, for us, is some discrete set of the form $\mathbb{Z} \times \mathbb{T}$ whose elements are known as sites. Before describing the spaces and operators at the sites we shall consider the lattice itself in some detail. For simplicity we shall always choose $\mathbb{T} = \mathbb{Z}^{\nu-1}$ so that the lattice is $\mathbb{Z}^\nu$, but the reader can easily supply the details for a more general oblique lattice. Moreover, that Ising model which we shall examine is constructed over $\mathbb{Z}^2$, i.e. $\nu = 2$.

Local regions are finite subsets of $\mathbb{Z}^\nu$, and the set of them is denoted by $\mathscr{L} = \{\Lambda\}$; we shall keep our usual abbreviation of $\mathscr{L}' = \mathscr{L} \cup \{\mathbb{Z}^\nu\}$. A lattice site, then, is a ν-tuple of integers $\mathbf{n} = (n_1, \ldots, n_\nu)$ with coordinatewise vector operations.

As distinguished local regions we choose

$$\Lambda_p = \{n \in \mathbb{Z}^\nu : |n_i| \leqslant p, i = 1, 2, \ldots, \nu\} \qquad (p \in \mathbb{N}); \tag{5.1}$$

the family $\mathscr{M} = \{\Lambda_p : p \in \mathbb{N}\}$ is our distinguished countable absorbing subfamily of $\mathscr{L}$, covering $\mathbb{Z}^\nu$:

$$\bigcup_{p \in \mathbb{N}} \Lambda_p = \mathbb{Z}^\nu.$$

Now the number of elements of (sites in) Λ_p is $2p + 1$. We shall write $|\Lambda|$ for the cardinality of $\Lambda \in \mathscr{L}$, and so $|\Lambda_p| = 2p + 1$; obviously $|\Lambda|$ corresponds to the volume of a local region in our $\mathbb{R}^3$-based models.

It is convenient to have a symbol for the family of those local regions with the same cardinality, so we write

$$\mathscr{L}_p = \{\Lambda \in \mathscr{L} : |\Lambda| = p\} \quad (p \in \mathbb{N}). \tag{5.2}$$

This completes our notation for the lattice itself. We next consider the spin spaces at the lattice sites.

For a spin-s system, $s \in \frac{1}{2}\mathbb{N} = \{\frac{1}{2}, 1, \frac{3}{2}, \ldots\}$, one associates a Hilbert space $\mathfrak{h}^{(\mathbf{n})} \simeq \mathbb{C}^{2s+1}$ to each site $\mathbf{n} \in \mathbb{Z}^\nu$. And to each local region $\Lambda \in \mathscr{L}$ one associates the $(2s+1)\,|\Lambda|$-dimensional Hilbert space

$$\mathfrak{h}(\Lambda) = \bigotimes_{\mathbf{n} \in \Lambda} \mathfrak{h}^{(\mathbf{n})}, \tag{5.3}$$

where, of course, the Hilbert tensor product is intended. (As $\mathfrak{h}(\Lambda)$ is finite-dimensional, no completion is necessary.)

In the previous chapter we had a similar construction to consider, but specialized to $s = \frac{1}{2}$. We saw there that in order to construct a systemic Hilbert space one had to choose a distinguished normalized vector $e_{\mathbf{n}} \in \mathfrak{h}^{(\mathbf{n})}$ for each $\mathbf{n}$ to act as a neutral vector. Once this family is chosen, one uses it to inject $\mathfrak{h}(\Lambda)$ into any $\mathfrak{h}(\Sigma)$ with $\Lambda \in \Sigma$, namely,

$$\begin{aligned} i(\Lambda, \Sigma) &: \mathfrak{h}(\Lambda) \to \mathfrak{h}(\Sigma); \\ i(\Lambda, \Sigma) &: \mathbf{v} \to \mathbf{v} \otimes (\bigotimes_{\Sigma \setminus \Lambda} e_n). \end{aligned} \tag{5.4}$$

Writing

$$\Omega = \bigotimes_{n \in \mathbb{Z}^\nu} e_n \tag{5.5}$$

for the unit vector distinguishing the incomplete direct product space, the now familiar inductive limit process gives the underlying Hilbert space for the lattice:

$$\begin{aligned} \mathscr{H}(\Omega) &= \varinjlim \{ i_{pq} [\ (\Lambda_p)]; (p,q) \in \mathbb{N}^2, p < q \} \\ &= \bigotimes^{\Omega}_{\mathbf{n} \in \mathbb{Z}^\nu} \mathfrak{h}^{(\mathbf{n})}. \end{aligned} \tag{5.6}$$

The construction of the spin algebras is also familiar. To each site $\mathbf{n} \in \mathbb{Z}^\nu$ we associate the algebra

$$\mathfrak{B}_n \equiv \mathbb{B}[\mathfrak{h}^{(\mathbf{n})}] \tag{5.7}$$

(which is vector space isomorphic to $\mathbb{C}^{(2s+1)^2}$). For each local region $\Lambda \in \mathscr{L}$ we choose

$$\mathfrak{B}(\Lambda) = \bigotimes_{\mathbf{n} \in \Lambda} \mathfrak{B}_{\mathbf{n}} \tag{5.8}$$

as the associated subalgebra; in particular, we write $\mathfrak{B}(\Lambda_p) \equiv \mathfrak{B}[p]$. The finite-dimensionality of $\mathfrak{B}_{\mathbf{n}}$ implies that $\mathfrak{B}(\Lambda)$ is given by the algebraic tensor product; no completion is necessary. Some topology is necessary for the quasilocal algebra, however. This is given through the C^*-inductive limit, or the norm closure of the unions of the local subalgebras. Eqns (2.37) and (2.38) give

$$\varphi_{nm} \;:\; \mathfrak{B}[n] \to \mathfrak{B}[m] \qquad (n \leqslant m), \tag{2.37a}$$

$$\varphi_{nm} \;:\; A \mapsto A \otimes \mathbf{1}_{m\setminus n} \qquad (A \in \mathfrak{B}[n]) \tag{2.37b}$$

for the requisite injective mappings, and so we may construct the quasilocal spin-$\frac{1}{2}$ algebra

$$\mathfrak{B} = \varinjlim\{\varphi_{nm}(\mathfrak{B}[n]; (m,n) \in \mathbb{N}^2, m > n\}. \tag{2.38}$$

Eqn (2.39), specialized to this lattice,

$$\varphi_n \;:\; \mathfrak{B}\,[n] \to \mathfrak{B},$$
$$\varphi_n(A) \;=\; A \otimes \mathbf{1}_{\mathbb{Z}\setminus n}\,, \tag{2.39a}$$

gives the injection of $\mathfrak{B}[n]$ into $\mathfrak{B}$. With it we can define what we shall call the local algebra for this lattice system:

$$\mathfrak{B}_{\mathrm{L}} \;=\; \bigcup_{p\in\mathbb{N}} \varphi_p\{\mathfrak{B}[p]\}, \tag{5.9a}$$

and then

$$\mathfrak{B} \;=\; \text{un. cl.}\,(\mathfrak{B}_{\mathrm{L}}). \tag{5.9b}$$

Lattice translations turn out to be important for the Ising model, and it is probably worth writing down the relevant formulae in this context. The operator

$$S^{(\mathrm{n})}(\mathrm{a}) \;:\; \mathfrak{h}^{(\mathrm{n})} \to \mathfrak{h}^{(\mathrm{n+a})} \qquad (\forall\, \mathrm{a} \in \mathbb{Z}^{\nu}) \tag{5.10}$$

is a one-one isometry, as each $\mathfrak{h}^{(\mathrm{m})} \simeq \mathbb{C}^{2s+1}$. The relations amongst these operators are

$$S^{(\mathrm{n})}(\mathrm{a}+\mathrm{b}) \;=\; S^{(\mathrm{n+b})}(\mathrm{a})\,S^{(\mathrm{n})}(\mathrm{b}) \qquad (\mathrm{a},\mathrm{b} \in \mathbb{Z}^{\nu}),$$
$$S^{(\mathrm{n})}(\mathbf{0}) \;=\; \mathbb{1}. \tag{5.11}$$

For any local region $\Lambda \in \mathscr{L}$ we define the translation operators

$$S_{\Lambda}(\mathrm{a}) \;=\; \bigotimes_{\mathrm{n}\in\Lambda} S^{(\mathrm{n})}(\mathrm{a}) \qquad (\forall\, \mathrm{a} \in \mathbb{Z}^{\nu}), \tag{5.12a}$$

whence

$$S_{\Lambda}(\mathrm{a}) \;:\; \mathfrak{h}(\Lambda) \to \mathfrak{h}(\Lambda + \mathrm{a}) \qquad (\forall\, \mathrm{a} \in \mathbb{Z}^{\nu}) \tag{5.12b}$$

univocally and isometrically.

Focusing our attention on the operators $S^{(p)}(\mathrm{a}) \equiv S_{\Lambda_p}(\mathrm{a})$, each such operator can be viewed as an element of $\mathfrak{B}$ if we identify it with $S_p(a) \otimes (\bigotimes_{\mathbb{N}\setminus p} \mathbf{1})$. The inductive limit structure of implies the existence of a limit operator as $p \to \infty$ which we write as $S(\mathrm{a})$. The family $\{S(\mathrm{a}) \;:\; \mathrm{a} \in \mathbb{Z}^{\nu}\}$ is the lattice-translation unitary group. One has that

$$S(\mathrm{a}) \;:\; \mathscr{H}(\Omega) \to \mathscr{H}(\Omega)$$
$$S(\mathrm{a}) \;;\; \bigotimes_{\mathbb{Z}^{\nu}} \nu_{\mathrm{n}} \to \bigotimes_{\mathbb{Z}^{\nu}} \nu_{\mathrm{n+a}} \qquad (\mathrm{a} \in \mathbb{Z}^{\nu}), \tag{5.13}$$

up to a rearrangement of the order of factors; one extends this to all of $\mathscr{H}(\Omega)$

by linearity and continuity. The group property follows from this:

$$S(\mathrm{a})\,S(\mathrm{b}) = S(\mathrm{a}+\mathrm{b}) \quad (\mathrm{a},\mathrm{b}\in \mathbb{Z}^\nu),$$
$$S(0) = 1; \tag{5.13a}$$

in the discrete topology on $\mathbb{Z}^\nu$, $S(\mathbb{Z}^\nu)$ is strongly continuous and unitary on $\mathscr{H}(\Omega)$.

The unitary group S (resp. the isometric endomorphisms S_Λ) implements the automorphism group

$$\sigma : \mathbb{Z}^\nu \to \mathrm{Aut}(\mathfrak{B}),$$
$$\sigma(\mathrm{a})[A] = S(\mathrm{a})\,A\,S(-\mathrm{a}) \tag{5.14}$$

(resp. the endormorphism family

$$\sigma_\Lambda : \mathbb{Z}^\nu \to \mathrm{End}[\mathfrak{B}(\Lambda), \mathfrak{B}(\Lambda+\mathrm{a})];$$
$$\sigma_\Lambda(\mathrm{a})[A] = S_\Lambda(\mathrm{a})\,A\,S_{\Lambda+\mathrm{a}}(-\mathrm{a})). \tag{5.15}$$

It is possible to prove that the system is asymptotically abelian with respect to space translations:

$$\lim_{|\mathrm{a}|\to\infty} \|[\sigma(\mathrm{a})[A], \mathbb{B}]_-\| = 0 \quad (A, B\in\mathfrak{B}), \tag{5.16}$$

$$|\mathrm{a}| = \left(\sum_1^\nu a_i^2\right)^{\frac{1}{2}}$$

is the Euclidean norm on $\mathbb{Z}^\nu$ (Robinson (2, 3)).

Before going on to consider the lattice dynamics we wish to make the following technical remarks concerning $\mathfrak{B}$, more or less as an aside. The algebra $\mathfrak{B}$ is very regular. As each $\mathfrak{B}_n$ is *simple* (no closed two-sided ideals), it follows (Sakai [1], proposition 1.23.8) that $\mathfrak{B}$ is simple; hence every non-trivial state is faithful (Emch [1], p.80).

Moreover $\mathfrak{B}$ is a *UHF (uniformly hyperfinite) algebra* (Sakai [1], definition 1.23.6 et seq.) and is *a forteriori* norm-separable. A UHF algebra is the C^*-inductive limit of Type-I_{p_n} factors with $p_n < \infty$ and $\lim_{n\to\infty} p_n = \infty$. This is true for $\mathfrak{B}$ with $p_n = (2s+1)(2n+1)^\nu$.

As for any C^*-algebra, the states $\mathfrak{S}(\mathfrak{B})$ on $\mathfrak{B}$ are normalized positive linear functionals. Note that, although $\mathfrak{B}$ is a tensor product of the algebras $\mathfrak{B}_n$, there are states on $\mathfrak{B}$ which are not simply tensor product states $\phi = \bigotimes_{\mathbb{Z}^\nu} \phi_n$, with $\phi_n \in \mathfrak{S}(\mathfrak{B}_n)$. For such a product state, note that if each ϕ_n is *factorial* (Sakai [1], definition 3.1.7), i.e. $\pi_{\phi_n}(\mathfrak{B}_n)''$ is a factor, then ϕ is factorial (Sakai [1], remark, p.75). Moreover, as each $\mathfrak{B}_n$ is finite-dimensional, each ϕ_n is primary, as is the product state ϕ.

5.3. Spin-lattice dynamics

Certain results are known for fairly general lattice systems; in this section we intend to point out some of them. As we are considering mechanistic systems, we assume some Hamiltonian $H(\Lambda)$ associated with each local region $\Lambda \in \mathbb{Z}^\nu$ to be given. We further assume it to be a self-adjoint (bounded) operator on $\mathfrak{h}(\Lambda)$ and to be translationally covariant:

$$S_\Lambda(\mathbf{a})\, H(\Lambda)\, S_{\Lambda+\mathbf{a}}(-\mathbf{a}) = H(\Lambda + \mathbf{a}); \tag{5.17}$$

when convenient we shall identify $H(\Lambda)$ with its image $H(\Lambda) \otimes \mathbf{1}_{\mathbb{Z}^\nu\setminus\Lambda}$ without further notation.

The general dynamical scheme was outlined in Chapter 1: $H(\Lambda)$ generates both the time-translation unitary group $U_\Lambda(t)$, $t \in \mathbb{R}$; and the Gibbs-state semi-group $\sigma_\Lambda(\beta)$, $\beta \in \mathbb{R}^+$. Thus

$$U_\Lambda(t) = \exp[\mathrm{i}tH(\Lambda)] \qquad (t \in \mathbb{R}), \tag{5.18a}$$

$$\sigma_\Lambda(\beta) = \exp[-\beta H(\Lambda)] \qquad (\beta \in \mathbb{R}^+), \tag{5.19a}$$

which leads to the time-translation automorphism group for the local subalgebras:

$$\tau_\Lambda : \mathbb{R} \to \mathrm{Aut}[\mathfrak{B}(\Lambda)] \qquad (\Lambda \in \mathscr{L}), \tag{5.18b}$$

$$\tau_\Lambda(t)[A] = U_\Lambda(t)\, A\, U_\Lambda(-t), \tag{5.18c}$$

and to the local Gibbs states

$$\phi_\Lambda^{(\beta)} \in \mathfrak{S}[\mathfrak{B}(\Lambda)] \qquad (\Lambda \in \mathscr{L}); \tag{5.19b}$$

$$\phi_\Lambda^{(\beta)}(A) = E_\Lambda^{(\beta)}(A) / E^{(\beta)}(1), \tag{5.19c}$$

with

$$E_\Lambda^{(\beta)}(A) = \mathrm{tr}_{\mathfrak{h}(\Lambda)}[\sigma_\Lambda(\beta) A]. \qquad (A \in \mathfrak{B}(\Lambda)). \tag{5.19d}$$

The existence of a time-translation automorphism group of $\mathfrak{B}$ generated locally by the τ_Λ depends critically upon the growth of $\|H(\Lambda)\|$ with site number $|\Lambda|$. We now consider this problem. Let us extend the definition eqn (2.37) of φ_{nm} to general local regions: $\varphi(\Lambda, \Sigma) : \mathfrak{B}(\Lambda) \to \mathfrak{B}(\Sigma)$, injecting $\mathfrak{B}(\Lambda)$ into $\mathfrak{B}(\Sigma)$ when $\Lambda \subset \Sigma$ by tensoring $A \in \mathfrak{B}(\Lambda)$ with the unit operator $\mathbf{1}_{\mathbb{Z}^\nu\setminus\Lambda}$. In this way φ_{nm} is identified with $\varphi(\Lambda_n, \Lambda_m)$. In addition we shall write $\varphi(\Lambda) : \mathfrak{B}(\Lambda) \to \mathfrak{B}$ for the identification of the local $\mathfrak{B}(\Lambda)$ with subalgebras of $\mathfrak{B}$. This injection is also effected by tensoring with the pertinent unit operator. We shall feel free to omit φ from formulae where the context makes it clear.

Considering the family $(H(\Lambda) : \Lambda \in \mathscr{L})$ of Hamiltonians as a whole, it is possible to replace it by a set theoretic function $\Phi : \mathscr{L} \to \mathfrak{B}$ in its stead. This *potential function* has the merit of being more closely related to the number of lattice sites which interact with one another than are the Hamiltonians. Indeed, the restriction of Φ to the family $\mathscr{L}_p$ of local regions containing p sites is known as the *p-body potential* for the model. The potential function Φ is defined recursively by the formula

$$H(\Lambda) = \sum_{\Delta \in \mathscr{P}(\Lambda)} \Phi(\Delta), \tag{5.20}$$

starting from $\Phi(\phi) = 0$. Here $\mathscr{P}(\Lambda)$ is the power set of Λ, of cardinality $2^{|\Lambda|}$.

For instance, $\Phi(\{\mathbf{n}\}) = H(\{\mathbf{n}\})$, $\Phi(\{\mathbf{n},\mathbf{m}\}) = H(\{\mathbf{n},\mathbf{m}\}) - H(\{\mathbf{n}\}) - H(\{\mathbf{m}\})$, and so on.

The *range* of Φ is a set $\rho \in \mathscr{L}'$ defined to be (recall the translation covariance eqn (5.17))

$$\rho = \{\mathbf{n} \in \mathbb{Z}^\nu : \text{there exists a } \Lambda \in \mathscr{L}' \text{ such that } \mathbf{n} \in \Lambda, \mathbf{o} \in \Lambda, \text{and } \Phi(\Lambda) \neq 0\}. \qquad (5.21)$$

Φ is of *finite range* if the range has a finite number of sites: $|\rho| < \infty$. It is customary to consider general classes of lattice models by starting from the set $\mathscr{P}_0$ of all finite range potentials and equipping it with some growth condition in the form of a norm, and then completing the resulting pre-Banach space.

Note: $\mathscr{P}_0$ is a vector space if we set $(\Phi + \Psi)(\Lambda) = \Phi(\Lambda) + \Psi(\Lambda)$ and $(c\Phi)(\Lambda) = c\Phi(\Lambda)$ as operators in $\mathfrak{B}$. We shall consider two norms on $\mathscr{P}_0$, $\|.\|_1$ and $\|.\|_2$, in what follows, writing $\mathscr{P}_1$ and $\mathscr{P}_2$ for the resulting Banach spaces. We define then

$$\|\Phi\|_1 = \sum_{\mathbb{Z}} \|\Phi(\Lambda)\|_{\mathfrak{B}} \exp(|\Lambda| - 1), \qquad (5.22a)$$

$$\|\Phi\|_2 = \sum_{\mathbb{Z}} \|\Phi(\Lambda)\|_{\mathfrak{B}}, \qquad (5.22b)$$

where $\|.\|_{\mathfrak{B}}$ is the $\mathfrak{B}$-norm, i.e. the operator norm on $\mathscr{H}(\Lambda)$.

One may prove that if $\Phi \in \mathscr{P}_1$ matters are extremely favourable. Robinson (1,2) has proved the main result.

THEOREM 5.1.

Let $\Phi \in \mathscr{P}_1$ and give rise to a Hamiltonian family $(H(\Lambda) : \Lambda \in \mathscr{L})$ which in turn generates the families $(U_\Lambda(t) = \exp[itH(\Lambda)] : \Lambda \in \mathscr{L}, t \in \mathbb{R})$ and $(\tau_\Lambda : \mathbb{R} \to \mathrm{Aut}[\mathfrak{B}(\Lambda)] : \Lambda \in \mathscr{L})$. Then there exists a quasilocal automorphism group $\tau : \mathbb{R} \to \mathrm{Aut}(\mathfrak{B})$ defined by

$$(\text{norm } \mathfrak{B}) - \lim_{n\to\infty} U_{\Lambda_n}(t)\, A\, U_{\Lambda_n}(-t) = \tau_t(A) \qquad (A \in \mathfrak{B}_{\mathrm{L}}) \qquad (5.23)$$

and extended from $\mathfrak{B}_{\mathrm{L}}$ to $\mathfrak{B}$ by continuity. Moreover, the automorphism group is strongly continuous, i.e.

$$\lim_{t\to s} \|\tau_t(A) - \tau_s(A)\| = 0 \qquad (A \in \mathfrak{B}).$$

For a proof, see Emch [1], theorem IV.2.2, Ruelle (1), §7.6, or Robinson (2,3).

For potentials in $\mathscr{P}_2$ things are more awkward. Ruskai (1) has proved the following theorem.

THEOREM 5.2.

Let $\Phi \in \mathscr{P}_2$, other notation as above, and let $\phi_n^{(\beta)} \equiv \phi_{\Lambda_n}^{(\beta)}$ be the Λ_n-region Gibbs state (eqn (5.19c)). *Then*

$$\lim_{n\to\infty} \phi_n^{(\beta)}(\tau_n(t_1)[A_1] \ldots \tau_n(t_k)[A_k])$$

exists for all $t_1, \ldots, t_k \in \mathbb{R}$ and all $A_1, \ldots, A_k \in \mathfrak{B}_{\mathrm{L}}$.

This being recognized as DS I, the general theory (Dubin and Sewell (1)) implies the following.

(1) A Gibbs state $\phi_\beta \in \mathfrak{S}(\mathfrak{B})$ exists and is defined by extension from

$$\phi_\beta(A) = \lim_{n\to\infty} \phi_n^{(\beta)}(A) \quad (A \in \mathfrak{B}_{\mathrm{L}}).$$

(2) The reconstruction by Wightman's method can be carried out. The resulting Hilbert space $\mathscr{H}_\beta$ contains a cyclic and separating vector Ω_β, whence there exists a modular automorphism group $\tau_\beta : \mathbb{R} \to \mathrm{Aut}[\pi_\beta(\mathfrak{B})'']$ unitarily implemented by a unitary group $U_\beta(\mathbb{R})$ generated by the modular operator (Takesaki [1]; Winnink (1)),

$$\begin{aligned}(\text{str. } \mathscr{H}_\beta)\text{-}\lim_{n\to\infty} \pi_\beta[\tau_n(t)A] &= \tau_\beta(t)\, \pi_\beta(A) \\ &= U_\beta(t)\, \pi_\beta(A)\, U_\beta(-t), \\ U_\beta(t)\, \Omega_\beta &= \Omega_\beta.\end{aligned}$$

(3) Writing Φ_β for the extension of ϕ_β to $\mathfrak{B}_\beta'' \equiv \pi_\beta(\mathfrak{B})''$, the vector state Φ_β is (β, τ_β)-KMS, whence it is locally normal (Takesaki and Winnink (1)); whence $\mathscr{H}_\beta$ is separable, implying that U_β is strongly continuous.

As DS II has not been proved, one does not know whether or not the GNS reconstruction space is a proper subspace of $\mathscr{H}_\beta$.

For $\mathscr{P}_1$ class potentials, DS I and II hold because of the norm convergence (5.22) of the automorphisms, so 1-3 holds *a forteriori* with the GNS reconstruction space equal to $\mathscr{H}_\beta$.

5.4. Spin-$\frac{1}{2}$ model dynamics

All the models we wish to discuss in this section are special cases of the Heisenberg magnet model. Using the symbol $J_{\mathbf{n}}^{(\alpha)}(m)$ for the αth spin operator at the point $\mathbf{n} \in \Lambda_m$, the model Hamiltonian is

$$H(p) = \sum_{\mathbf{n},\mathbf{m}\in\Lambda_p} \sum_{\alpha=1}^{3} \epsilon(\mathbf{n}, \mathbf{m}, \alpha)\, J_{\mathbf{n}}^{(\alpha)}(p)\, J_{\mathbf{m}}^{(\alpha)}(p) + \\ + \sum_{n\in\Lambda_p} b(\mathbf{n})\, J_{\mathbf{n}}^{(3)}(p). \tag{5.24}$$

Of course, we could generalize this formula to arbitrary $\Lambda \in \mathscr{L}$; and $\epsilon(\mathbf{n}, \mathbf{m}, \alpha)$ ought to indicate a Λ_p dependence. One might have the analogue of cyclic boundary conditions in mind, whence $\epsilon(\mathbf{n} + \mathbf{n}', \mathbf{m}, \alpha) = \epsilon(\mathbf{n}'', \mathbf{m}, \alpha)$, where $\mathbf{n} + \mathbf{n}' = \mathbf{n}''$ modulo Λ_p, for instance.

One varies the model by choosing an energy function $\epsilon: \Lambda_p \times \Lambda_p \times \{1,2,3\} \to \mathbb{R}$; and an external magnetic field $b : \Lambda_p \to \mathbb{R}$. If b is zero and ϵ is α-independent, the model is completely *isotropic*; if the values of ϵ are negative, the model exhibits *ferromagnetism* and conversely, positive ϵ implies *antiferromagnetism*. The translational covariance postulated for $H(p)$ in eqn (5.17) requires that the support of ϵ be the diagonal of $\Lambda_p \times \Lambda_p$, which is a technical way of saying that we shall take

$$\epsilon(\mathbf{n}, \mathbf{m}, \alpha) = \epsilon(\mathbf{n}-\mathbf{m}, \alpha) \tag{5.25}$$

hereafter. A special case much favoured for simplicity in computations is the *nearest-neighbour* Hamiltonian, for which $\epsilon(\mathbf{q}, \alpha)$ vanishes unless $\mathbf{q}$ is in the set

$$\{\pm\ (1,0,0,\ldots,0),\ \pm\ (0,1,0,\ldots,0),\ldots,\ \pm\ (0,0,\ldots,1)\},$$

where the elements are ν-tuples.

If $\epsilon(\mathbf{n}, \mathbf{m}, \alpha) = 0$ for $\alpha = 3$ and $b = 0$, the result is the XY-model, which is trivially solvable in the following sense. Proceeding heuristically, we can write the resulting XY-Hamiltonian in terms of fermion operators, and up to details it is quadratic and hence diagonalizable.

The other extreme is $\epsilon(\mathbf{n}, \mathbf{m}, \alpha) = \epsilon(\mathbf{n}-\mathbf{m})\,\delta_{\alpha\,3}$, whence

$$H(p) = \sum_{\mathbf{n},\mathbf{m}\in\Lambda_p} \epsilon(\mathbf{n}-\mathbf{m})\, J^{(3)}_{\mathbf{n}}(p)\, J^{(3)}_{\mathbf{m}}(p); \tag{5.26}$$

this defines the *Ising* Hamiltonian.

If we similarly restrict ourselves to the subalgebra of $\mathfrak{B}$ generated only by the $\{J^{(3)}_{\mathbf{m}} : \mathbf{m} \in \mathbb{Z}^\nu\}$ we have the usual Ising model; considering all of $\mathfrak{B}$ leads to the generalized Ising model. In spite of the fact that the Ising model is completely abelian, it is a very difficult problem to find the Gibbs state and the thermodynamic functions. For two dimensions, the thermodynamic functions were found by Onsager in 1939, although a full account was not published until several years later (Onsager (1)); the model for $\nu \geqslant 3$ remains unsolved.

Radin (1) has given an interesting computation for the time evolution in the generalized Ising model; we propose to state his results here. The particular Hamiltonian question has $b = 0$, $\epsilon(|\mathbf{n}-\mathbf{m}|)$ depending only upon the Euclidean distance $|\mathbf{p}| = [(p_1)^2 + \ldots + (p_\nu)^2]^{\frac{1}{2}}$, and in addition satisfying

$$\epsilon(0) = 0$$

$$\sum_{\mathbf{p}\in\mathbb{Z}^\nu} \epsilon(|\mathbf{p}|) < \infty. \tag{5.27}$$

These conditions ensure that there is a quasilocal automorphism $\tau : \mathbb{R} \to \mathrm{Aut}(\mathfrak{B})$ which is the norm limit of the local $\tau_n : \mathbb{R} \to \mathrm{Aut}(\mathfrak{B}[n])$ generated by the $U_n(t) = \exp[\mathrm{i}tH(n)]$:

$$\lim_{n\to\infty} \|\tau_n(t)A - \tau(t)A\| = 0 \qquad (A \in \mathfrak{B}_{\mathrm{L}}), \tag{5.28}$$

and extend to $\mathfrak{B}$ by continuity. The interest in this topic stems from the fact that Radin could find $\tau(t)A$ fairly explicitly. For two sets $\mathbf{A}, \mathbf{B}$ with $\mathbf{A} \subset \mathbf{B}$, $\mathbf{B}\backslash\mathbf{A} \equiv \{b \in \mathbf{B} : b \notin \mathbf{A}\}$ is the *relative compliment.* In particular, to each $\Lambda_n \in \mathscr{M}$ we associate a set $\Phi_n \subset \mathbb{Z}^\nu \times \mathbb{Z}^\nu$ by the rule

$$\Phi_n = (\mathbb{Z}^\nu \times \mathbb{Z}^\nu)\backslash\,[(\mathbb{Z}^\nu \times \Lambda_n) \times (\mathbb{Z}^\nu \times \Lambda_n)]. \tag{5.29}$$

With this notation, $\tau(\mathbb{R})$ is determined by the following action on $\mathfrak{B}_L$, extending to $\mathfrak{B}$ by continuity. Since for every $A^{(n)} \in \mathfrak{B}_L$ there is an integer n such that $A^{(n)} \in \mathfrak{B}[m]$ for $m \geqslant n$ (up to identification) it suffices to consider τ on $\mathfrak{B}[n]$ for arbitrary n. Then

$$\tau(t)\,A^{(n)} = V^{(n)}(t)\,A^{(n)}\,V^{(n)}(-t) \qquad (\forall A^{(n)} \in \mathfrak{B}[n]), \tag{5.30a}$$

with $V^{(n)}(\mathbb{R})$ the strongly continuous unitary group

$$V^{(n)}(t) = \exp\left\{\mathrm{i}t \sum_{(\mathbf{p},\mathbf{q})\in\Phi_n} \epsilon(|\mathbf{p}-\mathbf{q}|)\,J^{(3)}_{\mathbf{p}}\,J^{(3)}_{\mathbf{q}}\right\}. \tag{5.30b}$$

We now apply this formula to the generators. Let $\omega_{\mathbf{p}} \in \mathfrak{B}$ be the 'frequency' operator

$$\omega_{\mathbf{p}} = \sum_{\mathbf{q}\in\mathbb{Z}^\nu} \epsilon(|\mathbf{p}-\mathbf{q}|)\,J^{(3)}_{\mathbf{q}}, \tag{5.31}$$

then

$$\tau(t)\,J^{(3)}_{\mathbf{p}} = J^{(3)}_{\mathbf{p}}, \tag{5.32a}$$

$$\tau(t)\,J^{(1)}_{\mathbf{p}} = J^{(1)}_{\mathbf{p}}\cos(\omega_{\mathbf{p}}t) - J^{(2)}_{\mathbf{p}}\sin(\omega_{\mathbf{p}}t), \tag{5.32b}$$

$$\tau(t)\,J^{(2)}_{\mathbf{p}} = J^{(1)}_{\mathbf{p}}\sin(\omega_{\mathbf{p}}t) + J^{(2)}_{\mathbf{p}}\cos(\omega_{\mathbf{p}}t), \tag{5.32c}$$

or

$$\tau(t)\,J^{(+)}_{\mathbf{p}} = J^{(+)}_{\mathbf{p}}\exp(+\,\mathrm{i}t\omega_{\mathbf{p}}). \tag{5.32d}$$

Before going on to consider the Ising model in detail, we mention a result of Robinson, who proved the existence of a phase transition in the isotropic Heisenberg model. In this case one considers only nearest-neighbour interactions: $\epsilon(\mathbf{p},\alpha) = 0$ unless $\mathbf{p} = \mathbf{e}_i = (\delta_{ij} : j = 1, \ldots, \nu)$, for $i = 1, \ldots, \nu$; and $b(\mathbf{n}) = b$ independent of $\mathbf{n}$. Writing

$$\lambda_\alpha = 4 \sum_{i=1}^{\nu} |\epsilon(\mathbf{e}_i, \alpha)|, \tag{5.33}$$

Robinson (3) showed that a phase transition occurs for large enough β and small enough λ_1/λ_3 and λ_2/λ_3; it is assumed that $\epsilon(\mathbf{e}_i, 3) \neq 0$. There are two phases and a non-zero spontaneous magnetization. The method used to prove this result, which we will not do, is a direct descendant of Peierl's early work on the Ising model. The reader is referred to Robinson's paper and to more recent papers on the Ising model for a description of these graph-theoretic techniques used to obtain estimates of pertinent operator bounds; cf. Emch [1] for references.

5.5. Ising model kinematics

The Ising model is justly famous for being the only exactly solvable model ($\nu = 2$) showing a phase transition supporting a spontaneous magnetization. In order to solve the model, the ν-dimensional abelian model is mapped onto a $(\nu - 1)$-dimensional full spin-$\frac{1}{2}$ model, i.e. a system $\mathfrak{B}$ as above, for for $(\nu - 1)$. Clearly this is an advantage only for $\nu = 2$. One also see that it is as difficult to solve the ν-dimensional Ising model problem as the $(\nu - 1)$-dimensional Heisenberg model problem.

As a peripheral matter, let us mention a connection with quantum field theory. The conjecture of Schwinger (1) and Symanzik (1) that a Euclidean invariant quantum field theory ought to be equivalent to a relativistic theory has now been shown to be true under certain reasonable restrictions (e.g. Osterwalder and Schrader (1)). Restriction of the attention to self-interacting bosons, a technique much like the transfer matrix method used to solve the Ising model, proves the equivalence of a four-dimensional self-interacting boson field theory, say a $(\varphi)^p_4$ theory, with a five-dimensional Ising-type theory but continuous (i.e. over $\mathbf{R}^5$ rather than $\mathbb{Z}^5$). If one could prove an approximation theorem for the existence of the limit Ising $(\mathbf{R}^5) \leftarrow$ Ising $(\mathbb{Z}^5)$, one could then have a proof of the existence of $(\varphi)^p_4$, which would be an important result (Nelson (1, 2); Guerra, Rosen, and Simon (1)).

Our purpose in examining this model ($\nu = 2$) essentially is to discover its general algebraic structure, but unfortunately the computations are quite long and involved. The two standard textbook references for the necessary calculations are Huang [1] (after Kaufmann (1)) and the approach of Schultz, Mattis, and Lieb (1) described in Mattis [1]. We shall follow a simplified version of the work of Marinaro and Sewell (1). The main simplification arises in failing to provide both detail and proof.

Our first task is to consider the restricted algebra associated with the $\{J^{(3)}_{\mathbf{p}}\}$ only. Hereafter, we shall write $\mathscr{L}_2$ and $\mathscr{L}'_2$ for $\mathscr{L}$ and $\mathscr{L}'$ constructed over $\mathbb{Z}^2 = \mathbb{Z} \times \mathbb{Z}$ and $\mathscr{L}_1$ (resp. $\mathscr{L}'_1$ for the construction over $\mathbb{Z}$).

Let us write $[N]$ for the $(2N + 1)$-point set

$$[N] = \{0, \pm 1, \ldots, \pm N\} \subset \mathbb{Z}; \tag{5.34}$$

such sets constitute the absorbing cover $\mathscr{N} = \{[N] : N \in \mathbf{N}\}$ of $\mathbb{Z}$ which we shall use. And making use of the rectangularity of $\mathbb{Z}^2 = \mathbb{Z} \times \mathbb{Z}$ we shall choose $\mathscr{M} = \mathscr{N} \times \mathscr{N} = \{[N] \times [P]\}$ for the $\mathbb{Z}^2$ cover.

The Ising model algebras are subalgebras of the full lattice algebras, and we write, for every $\Lambda \in \mathscr{L}_2$,†

$$\mathscr{I}(\Lambda) = \langle 1(\Lambda), J^{(3)}_n(\Lambda) : n \in \Lambda \rangle, \tag{5.35}$$

which is finite-dimensional and so is *a forteriori* closed: $\mathscr{I}(\Lambda) \subset \mathfrak{B}(\Lambda; \nu = 2)$. The abbreviation $\mathscr{I}_{nm} \equiv \mathscr{I}([n] \times [m])$ is convenient, for up to obvious identifications, the quasilocal $\mathbb{Z}^2$ Ising model algebra is the C^*-inductive limit algebra

$$\mathscr{I} = \varinjlim \{\mathscr{I}_{nm} : (n, m) \in \mathbf{N}^2\}. \tag{5.36}$$

Clearly $\mathscr{I} \subset \mathfrak{B}(\nu = 2)$ is a C^*-subalgebra of $\mathfrak{B}(\nu = 2)$.

† Recall that $\langle \cdot \rangle$ means the algebraic span in the indicated variables.

For any $\Delta \in \mathscr{L}_1$, the *n*th-*row algebra* is defined to be

$$\mathscr{R}_n(\Delta) = \mathscr{I}(\{n\} \times \Delta); \tag{5.37a}$$

for $\Delta = \mathbb{Z}$ we mean the partial C^*-inductive limit

$$\begin{aligned} \mathscr{R}_n(\mathbb{Z}) &\equiv \mathscr{R}_n \\ &= \lim_{p \to \infty} \{\mathscr{I}(\{n\} \times [p])\}. \end{aligned} \tag{5.37b}$$

We may observe here that the C^*-inductive limit $\lim_{n \to \infty} \mathscr{R}_n = \mathscr{I}$. These row algebras will be used later.

A general structure theorem for C^*-algebras relates abelian C^*-algebras to function spaces of the form $\mathscr{C}(\mathscr{S})$ for certain spaces $\mathscr{S}$ which depend upon the algebra in question. We shall now construct such a relation for the Ising algebras $\mathscr{I}(\Lambda)$ and $\mathscr{R}_n(\Lambda)$.

Let $\mathscr{T}$ be the two-point set

$$\mathscr{T} = \{-1, +1\}. \tag{5.38}$$

By $\mathscr{C}(\mathscr{T})$ we mean the Banach space of continuous functions $\{f : \mathscr{T} \to \mathbb{C}\}$ equipped with the *sup*-norm:

$$\|f\| = \sup\{|f(-1)|, |f(+1)|\}.$$

By $\mathfrak{r}$ we shall mean the 2×2 matrix algebra $\left\{\begin{bmatrix} a & 0 \\ 0 & b \end{bmatrix} : a, b \in \mathbb{C}\right\}$. It is easy to check that the mapping

$$\begin{bmatrix} a & 0 \\ 0 & b \end{bmatrix} \to (f(+1) = a, f(-1) = b)$$

defines a C^*-isomorphism between $\mathfrak{r}$ and $\mathscr{C}(\mathscr{T})$. This is a special case of a general structure theorem which we shall now apply to the Ising model (Gel'fand, Raikov, and Shilov [1]).

Equipping $\mathscr{T}$ with the discrete topology (every subset of $\mathscr{T}$ is open), it becomes a compact Hausdorff space. For any $\Lambda \in \mathscr{L}_j$ (j either 1 or 2), let $\mathfrak{T}^\Lambda$ be the topological product $\times_{\lambda \in \Lambda} \mathscr{T}$ (Choquet [1]); by Tychonov's theorem (Choquet [1]), $\mathscr{T}^\Lambda$ is compact. A standard result in analysis is the tensor product result $\bigotimes_{\lambda \in \Lambda} [\mathscr{C}(\mathscr{T})]_\lambda = \mathscr{C}(\mathscr{T}^\Lambda)$; no problem of completion arises since $|\Lambda| < \infty$.

This result gives the required C^*-isomorphisms upon specializing the choice of Λ and using $\mathfrak{r} \simeq \mathscr{C}(\mathscr{T})$, namely,

$$\mathscr{I}(\Lambda) \simeq \mathscr{C}(\mathscr{T}^\Lambda) \qquad (\Lambda \in \mathscr{L}_2) \tag{5.39}$$

and

$$\mathscr{R}_n(\Delta) \simeq \mathscr{C}(\mathscr{T}^{\{n\} \times \Delta}) \qquad (\Delta \in \mathscr{L}_1). \tag{5.40a}$$

It is convenient to write

$$\mathscr{C}(\mathscr{T}^\Delta)_n \equiv \mathscr{C}(\mathscr{T}^{\{n\} \times \Delta}). \tag{5.40b}$$

We can be even more explicit. The correspondence between $\mathscr{I}(\Lambda)$ and $\mathscr{C}(\mathscr{T}^\Lambda)$ is defined by the formula

$$J_n^{(3)}(\Lambda) \to (\delta_n : t \mapsto t_n), \tag{5.39a}$$

where t_n is the nth projection of $t = (t_p)_{p \in \Lambda} \in \mathscr{C}(\mathscr{T}^\Lambda)$.

For $\mathscr{R}_n(\Delta) \simeq \mathscr{C}(\mathscr{T}^\Delta)_n$ we have a similar formula: specializing eqn (5.39a) to $\Lambda = \{n\} \times \Delta$, we have

$$J_{(nm)}^{(3)}(\{n\} \times \Delta) \to (\delta_m^{(n)} : t \mapsto t_{(nm)}). \tag{5.40c}$$

The C^*-isomorphism to the nth row algebra arises from identifying this generating element with

$$J_m^{(3)[n]}(\Delta) \to (\delta_m^{(n)} : t \mapsto t_{(nm)}) \quad (m \in \Delta). \tag{5.40d}$$

5.6. Ising model dynamics and the transfer matrix

Let us restrict our attention to the simplified Hamiltonian $H_{nm} = H([n] \times [m])$ given by

$$\begin{aligned} H_{nm} &= -g_1 \sum_{j=-n}^{+n} \sum_{k=-m}^{+m} J_{(jk)}^{(3)} J_{(j\,k+1)}^{(3)} - \\ &\quad - g_2 \sum_{j=-n}^{+n} \sum_{k=-m}^{+m} J_{(jk)}^{(3)} J_{(j+1\,k)}^{(3)} \end{aligned} \tag{5.41}$$

with cyclic boundary conditions $n + 1 \equiv -n, m + 1 \equiv -m$. Breaking up the sums this way, we have separated the energy of interaction along each row (first term) from that between neighbouring rows (second term).

Going over to function-space notation, we have the Hamiltonian function $\hat{H}_{nm} \in \mathscr{C}(\mathscr{T}^{[n] \times [m]})$ corresponding to the Hamiltonian operator H_{nm}:

$$\begin{aligned} \hat{H}_{nm}(t) &= -g_1 \sum_{j=-n}^{+n} \sum_{k=-m}^{+m} t_{(jk)} t_{(j\,k+1)} - \\ &\quad - g_2 \sum_{j=-n}^{+n} \sum_{k=-m}^{+m} t_{(jk)} t_{(j\mp 1\,k)}, \end{aligned} \tag{5.42}$$

with $n + 1 \equiv -n, m + 1 \equiv m$, and $t = (t_{(jk)}) \in \mathscr{T}^{[n] \times [m]}$.

Once we have $\hat{H}_{nm}$ we can write the formal expression for the (n, m)-Gibbs state $\phi_{nm}^{(\beta)}$ in function space form. For every function $\hat{A} \in \mathscr{C}(\mathscr{T}^{[n] \times [m]})$, we set

$$E_{nm}^{(\beta)}(\hat{A}) = \sum_{t \in \mathscr{T}^{[n] \times [m]}} \hat{A}(t) \exp[-\beta H_{nm}(t)], \tag{5.43a}$$

and then

$$\phi_{nm}^{(\beta)}(A) = E_{nm}^{(\beta)}(A)/E_{nm}^{(\beta)}(1). \tag{5.43b}$$

Our notation is intended to correlate $A \in \mathscr{I}(\Lambda)$ with $\hat{A} \in \mathscr{C}(\mathscr{T}^\Lambda)$.

It will suffice to consider row-product operators as arguments of $\phi_{nm}^{(\beta)}$. Writing $A^{(n)}$ for a typical operator in $\mathscr{R}_n([m])$, a row-product operator in $\mathscr{I}_{nm}$ is an element of the form

$$A = \bigotimes_{j=-n}^{+n} A^{(j)} \quad (A^{(j)} \in \mathscr{R}_j([m])), \tag{5.44a}$$

since $\mathscr{I}_{nm} = \bigotimes_{j=-n}^{+n} \mathscr{R}_j([m])$. To find the corresponding function $A \in \mathscr{C}(\mathscr{T}^{[n] \times [m]})$

we decompose $t \in T^{[n]\times[m]}$ by rows: for each $j \in [n]$, $t^{(j)} = (t_{jk})_{k\in[m]}$ is in $\mathscr{T}^{\{j\}\times[m]}$ and then $(t^{(j)})_{j\in[n]}$ is in $\mathscr{T}^{[n]\times[m]}$. That is, $t^{(j)}$ is the jth-row component of t. If $A^{(j)} \in \mathscr{C}(\mathscr{T}^{\{j\}\times[m]})$ is the function image of $A^{(j)} \in \mathscr{R}_j([m])$, the image of A of (5.44a) is

$$\hat{A}(t) = \prod_{j=-n}^{+n} \hat{A}^{(j)}(t^{(j)}). \tag{5.44b}$$

After a calculation which can be found carefully set out in Huang [1], one finds that

$$\begin{aligned} E_{nm}^{(\beta)}(\hat{A}) = \sum_{t\in\mathscr{T}^{[n]\times[m]}} & \prod_{j=-n}^{n_1} \hat{V}_m[t^{(j)}, t^{(j+1)}]\,\hat{A}^{(n_1)}(t^{n_1}) \times \\ & \times \prod_{j=n_1}^{n_2-1} \hat{V}_m[t^{(j)}, t^{(j+1)}]\,\hat{A}^{(n_2)}(t^{(n_2)}) \times \ldots \times \\ & \times \hat{A}^{(n_k)}(t^{(n_k)}) \prod_{j=n_k}^{+n} \hat{V}_m[t^{(j)}, t^{(j+1)}]. \end{aligned} \tag{5.45}$$

In this complicated formula, the indices $n_1, n_2, \ldots, n_k \in [n]$ correspond to just those j in the product in eqn (5.44a) for which $A^{(j)}$ is not a unit operator **1**. The function $\hat{V}_m \in \mathscr{C}(\mathscr{T}^{[m]\times[m]})$ is the 'transfer matrix' function, where $t^{(j)} \in \mathscr{T}^{\{j\}\times[m]}$ is identified with an element of $\mathscr{T}^{[m]}$ by an abuse of notation. The actual formula defining $\hat{V}_m$ is

$$\begin{aligned} \hat{V}_m[t^{(j)}, t^{(j+1)}] = {} & \exp\left[\frac{\beta g_1}{2} \sum_{k=-m}^{+m} t_k^{(j)} t_{k+1}^{(j)}\right] \times \\ & \times \exp\left[-\frac{\beta g_2}{2} \sum_{k=-m}^{+m} t_k^{(j)} t_k^{(j+1)}\right] \times \\ & \times \exp\left[\frac{\beta g_1}{2} \sum_{k=-m}^{+m} t_k^{(j+1)} t_k^{(j+1)}\right], \end{aligned} \tag{5.46}$$

with cyclic boundary conditions: $n+1 \equiv -n$, $m+1 \equiv -m$.

The key to the solution of evaluating $E_{nm}^{(\beta)}$—and this evaluation is not as explicit in this model as has been the case for the previous models—lies in replacing this two-dimensional abelian problem by a one-dimensional non-commutative problem. Recalling eqns (5.39)-(5.40) we relate $\mathfrak{B}([m])$ to $\mathscr{C}(\mathscr{T}^{[m]} \times \mathscr{T}^{[m]})$ (for each fixed row j) as follows:

$$\begin{aligned} J_m^{(1)} &\mapsto \delta_{(jm)}^{(1)}[t] = \tfrac{1}{2}(t_m^{(j)} + t_m^{(j+1)}), \\ J_m^{(2)} &\mapsto \delta_{(jm)}^{(2)}[t] = \tfrac{i}{2}(t_m^{(j)} - t_m^{(j+1)}), \\ J_m^{(3)} &\mapsto \delta_{(jm)}^{(3)}[t] = \tfrac{1}{2}(1 - t_m^{(j)} t_m^{(j+1)}). \end{aligned} \tag{5.47}$$

Using this correspondence we can rewrite $\hat{V}_m$ in terms of these $J_m^{(\alpha)}$ operators; the resulting expression may be viewed as an operator in $\mathfrak{B}([m])$. And by viewing the row algebras $(\mathscr{R}_j[m])_{j\in[n]}$ as subalgebras of $\mathfrak{B}([m])$, eqn (5.45) for $E_{nm}^{(\beta)}(A)$ can be rewritten in these terms. Let us write $V_m \in \mathfrak{B}([m])$ for the rewritten transfer matrix. Then (cf. Marinaro and Sewell (1))

$$E^{(\beta)}_{nm}(A) = \mathrm{tr}[A^{(n_1)}(V_m)^{n_2-n_1}A^{(n_2)}(V_m)^{(n_3-n_2)}A^{(n_k)} \times$$
$$\times\ (V_m)^{2n+n_1-n_k+1}] \tag{5.48}$$

with the trace over the Hilbert space $\bigotimes_{j=-m}^{+m}(\mathbb{C}^2)_j = \mathfrak{h}([m])$.

5.7. The global Gibbs state

We now consider the thermodynamic limit for this model. The operator V_m is a positive bounded self-adjoint invertible operator on the finite-dimensional space $\mathfrak{h}([m])$. By the Perron-Frobenius theorem (Gantmacher [1]) it follows that the largest eigenvalue λ_m of V_m is non-degenerate; we write Ω_m for the unique normalized principal eigenvector: $V_m\Omega_m = \lambda_m\Omega_m$, $\|\Omega_m\| = 1$. The form of $E^{(\beta)}_{nm}(A)$ in eqn (5.48) shows that the combination $V_mBV_m^{-1}$ for $B \in \mathfrak{B}([m])$ is clearly important, so we define the automorphism

$$\begin{aligned} w_m &\in \mathrm{Aut}\{\mathfrak{B}([m])\}; \\ w_m(B) &= V_mBV_m^{-1}. \end{aligned} \tag{5.49}$$

One might call this the *transfer automorphism.* We also write $\omega_m \in \mathfrak{S}\{\mathfrak{B}([m])\}$ for the vector state associated with the principal eigenvector:

$$\omega_m(B) = \langle\Omega_m, B\Omega_m\rangle. \tag{5.50}$$

Marinaro and Sewell (1) show that

$$\lim_{n\to\infty}\phi^{(\beta)}_{nm}(A) = \omega_m\{A^{(n_1)}w_m^{(n_2-n_1)}[A]^{(n_2)}\ldots w^{(n_k-n_1)}[A]^{(n_k)}\}, \tag{5.51}$$

where $w_m^{(j)}[B] = V_m^jB(V_m^{-1})^j$ is the j-fold composition of w_m with itself.

Assuming this result, it remains to take the limit $m\to\infty$. Since $\mathfrak{B}([m])$ is finite, the limit $\lim_{m\to\infty}\omega_m = \omega$ exists as a state on $\mathfrak{B} = \varinjlim\mathfrak{B}([m])$, but the Perron-Frobenius theorem does not apply to it. We write

$$\begin{aligned}\lim_{m\to\infty}\omega_m(B) &= \omega(B) \\ &= (\Omega, \pi(B)\Omega) \quad (B\in\mathfrak{B}_{\mathrm{L}}),\end{aligned} \tag{5.52}$$

where $\omega \sim [\mathscr{H}, \pi, \Omega]$ is the GNS association. Note that ω may well depend upon the particular $\mathbb{Z}^2$ cover $\mathscr{M} = \mathscr{N}\times\mathscr{N}$ chosen.

Marinaro and Sewell (1) also show that the automorphisms $\{w_m\}$ converge in the sense that there exists a positive self-adjoint contraction operator $V \in \mathbb{B}(\mathscr{H})$, $\|V\| \leqslant 1$, defined by

$$\lim_{m\to\infty}\pi[w_m(B)]\Omega = V\pi(B)\Omega \quad (B\in\mathfrak{B}_{\mathrm{L}}), \tag{5.53a}$$
$$V\Omega = \Omega. \tag{5.53b}$$

The final quadrature for the Ising model quasilocal Gibbs state is

$$\phi_\beta(A) = (\Omega, \pi(A)^{n_1} V^{n_2-n_1} \pi(A)^{n_2} V^{n_3-n_2} \ldots \pi(A)^{n_k} \Omega). \tag{5.54}$$

5.8. Interlude: one dimension

In one dimension, $\nu = 1$, the corresponding calculation is trivial, but no phase transition occurs. Taking $g_1 = g_2 = g$ and noting that there is only one row, the transfer matrix is

$$\acute{V} = \begin{bmatrix} \exp(\beta g) & \exp(-\beta g) \\ \exp(-\beta g) & \exp(\beta g) \end{bmatrix}. \tag{5.55}$$

Its eigenvalues are $\lambda_+ = \cosh(\beta g)$ and $\lambda_- = \sinh(\beta g)$; we introduce the ratio $\xi = \lambda_-/\lambda_+ = \tanh(\beta g)$, so that $0 < \xi < 1$.

For $p_1 < p_2 \ldots < p_{2n}$ and all these $p_j \in [P]$, we have the $[P]$-region Gibbs state:

$$\phi_P^{(\beta)}(J_{p_1}^{(3)} \ldots J_{p_{2n}}^{(3)}) = [\lambda_+^P + \lambda_-^P]^{-1} [\xi^{-p_1+p_2 \cdots + p_{2n}}) \lambda_+^P + \\ + (\xi^{p_1-p_2 \cdots p_{2n}}) \lambda_-^P]. \tag{5.56a}$$

Its limit is easily found:

$$\phi_\beta(J_{p_1}^{(3)} \ldots J_{p_{2n}}^{(3)}) = \xi^{-p_1+p_2- \cdots + p_{2n}}. \tag{5.56b}$$

We note that ϕ_β is a product state and has no associated phase transition.

5.9. Spontaneous breakdown of spin-reversal symmetry

Returning to the two-dimensional problem we shall describe some properties of the quasilocal Gibbs state ϕ_β. The phase transition is mathematically due to the bifurcation of the principal eigenvalue in the limit; and this bifurcation is associated with the spontaneous breakdown of spin-reversal symmetry. This symmetry is given by the automorphism $\rho \in \mathrm{Aut}(\mathscr{J})$ defined by its action on generators:

$$\rho : \mathbf{1} \mapsto \mathbf{1}, \\ \rho : J_{\mathbf{n}}^{(3)} \mapsto -J_{\mathbf{n}}^{(3)} \quad (\mathbf{n} \in \mathbb{Z}^2). \tag{5.57}$$

We shall also need the row-translation symmetry group

$$\sigma : \mathbb{Z} \to \mathrm{Aut}(\mathscr{J}); \\ \sigma_n[\mathbf{1}] = \mathbf{1} \\ \sigma_n[J_{(mp)}^{(3)}] = J_{(m+np)}^{(3)}. \tag{5.58}$$

With this notation, the basic structural theorem for the $\mathbb{Z}^2$-Ising model under consideration here is as follows.

THEOREM 5.3.

There exists a state $\phi_\beta \in \mathfrak{S}(\mathscr{I})$, defined by extension from

$$\lim_{\mathscr{M}=\mathscr{N}\times\mathscr{N}} \phi^\beta_{(mn)}(A) = \phi_\beta(A) \quad (A \in \mathscr{I}_\mathrm{L}), \tag{5.59}$$

which is $\sigma(\mathbb{Z})$-invariant and ρ-invariant. There exists a critical temperature β_c such that for $\beta < \beta_c$, ϕ_β is $\phi(\mathbb{Z})$-ergodic. For $\beta \geqslant \beta_c$, ϕ_β undergoes a ρ-symmetry break-down associated with its unique $\sigma(\mathbb{Z})$-ergodic decomposition

$$\phi_\beta = \int^{\oplus}_{\mathfrak{E}[\mathscr{I},\sigma(\mathbb{Z})]} \Psi_\beta \, d\nu(\Psi_\beta), \tag{5.60}$$

$\Psi_\beta \notin \mathfrak{S}(\mathscr{I}, \rho)$ (ν-almost all).

For a proof see Marinaro and Sewell (1). In one way and another this theorem is due to the work of many people, certain of who have already been mentioned. Further bibliography will be found in the cited references.

Emch, Knops, and Verboven (2) have shown how the two-fold degeneracy appears in this context. If one considers the truncated problem of only observables from one row algebra, say $\mathscr{R}_n$, the corresponding Gibbs state is the restriction $\phi_{\beta|n}$ of our limit state ϕ_β. One finds that $\phi_{\beta|n} = \frac{1}{2}(\omega_n^+ + \omega_n^-)$, where $\omega_n^\pm$ are vector states which are ρ-extremal and satisfy $\rho^*(\omega_n^\pm) = \pm\, \omega_n^\pm$.

Bibliography

Articles

ABRAHAM, D. B., GALLAVOTTI, G., and MARTIN-LÖF, A. (1973) (1) Surface tension on the two-dimensional Ising model. *Physica* **65**.

and MARTIN-LÖF, A. (1973) (1) The transfer matrix for a pure phase in the two-dimensional Ising model. *Commun. math. Phys.* **32**.

ARAKI, H. (1961-2) (1) *Einführung in die axiomatische Quantenfeld Theorie.* Lecture Notes of Eidgenössiche Technische Hochschule, Zürich. (1963) (2) A lattice of Von Neumann algebras associated with the quantum theory of a free bose field. *J. math. Phys.* **4**. (1964). (3) Von Neumann algebras of local observables for a free scalar field. *J. math. Phys.* **5**. (1968) (4) Multiple time analyticity of a quantum statistical state satisfying the KMS boundary condition. *Publ. Res. Inst. Math. Sci. Kyoto Univ.*, Ser. A **4**. (1969) (5) Gibbs states of a one dimensional quantum lattice. *Commun. math. Phys.* **14**.

and MIYATA, H. (1968) (1) On KMS boundary condition. *Publ. Res. Inst. Math. Sci. Kyoto Univ.* Ser. A, **4**.

and WOODS, E. J. (1963) (1) Representations of the canonical commutation relations describing a nonrelativistic infinite free bose gas *J. math. Phys.*, **4**.

and WYSS, W. (1964) (1) Representations of canonical anticommutation relations. *Helv. phys. Acta* **37**.

BARDEEN, J., COOPER, L. N., and SCHRIEFFER, J. R. (1957) (1) Theory of superconductivity. *Phys. Rev.* **108**.

BAUMANN, K., EDER, G., SEXL, R., and THIRRING, W. (1961) (1) On field theories with degenerate ground states. *Ann. Phys.* **16**.

BOGOLIUBOV, N. N. (1958) (1) A new method in the theory of superconductivity. *Soviet Phys. JETP* **7**.

BOSE, S. N. (1924) (1) Planck's Gesetz und Lichtquantenhypothese. *Z. Phys.* **26**.

BRASCAMP, H. J. (1970) (1) Equilibrium states for a classical lattice gas. *Commun. math. Phys.* **18**.

CANNON, J. T. (1973) (1) Infinite volume limits of the canonical free bose gas states on the Weyl algebra. *Commun. math. Phys.* **29**.

CHAIKEN, J. M. (1967) (1) Finite particle representations and states of the canonical commutation relations. *Ann. Phys.* **42**. (1968) (2) Number operators for representations of the canonical commutation relations. *Commun. math. Phys.* **8**.

COOK, J. M. (1953) (1) The mathematics of second quantization. *Trans Am. math. Soc.* **74**.

COOPER, L. N. (1956) (1) Bound electron pairs in degenerate Fermi gas. *Phys. Rev.* **104**.

DAVIES, E. B. (1972) (1) Diffusion for weakly coupled quantum oscillators.

Commun. Math. Phys. **27.** (1973) (2) The harmonic oscillator in a heat bath. *Commun. math. Phys.* **33.** (1973) (3) Dynamics of an infinite atom Dicke maser model. *Commun. math. phys.* **33.**

DELL'ANTONIO, G.-F. (1968) (1) Structure of the algebra of some free systems. *Commun. Math. Phys.* **9.** (1971) (2) Can local gauge transformations be implemented? *J. math. Phys.* **12.**

and DOPLICHER, S. (1967) (1) Total number of particles and Fock representation. *J. math. Phys.* **8.**

and RUELLE, D. (1966) (1) A theorem on canonical commutation and anticommutation relations. *Commun. Math. Phys.* **2.**

DIRAC, P. A. M. (1926) (1) On the theory of quantum mechanics. *Proc. R. Soc.* A **112.** (1928) (2) The quantum theory of the electron. *Proc. R. Soc.* A **117, 118.**

DOPLICHER, S., KADISON, R. V., KASTLER, D., and ROBINSON, D. W. (1967) (1) Asymptotically abelian systems. *Commun. math. Phys.* **6.**

and KASTLER, D. (1965) (1) Ergodic states in a noncommutative ergodic theory. *Commun. math. Phys.* **7.**

and STØRMER, E. (1969) (1) Invariant states and asymptotic abelianess *J. funct. Anal.* **3.**

DOPLICHER, S. and POWERS, R. T. (1968) (1) On the simplicity of the even CAR algebra and free field models. *Commun. Math. Phys.* **7.**

DUBIN, D. A. (1) Fock space formulation of the Heisenberg ferromagnet. *Nuovo Cim.* **53** B.

and SEWELL, G. L. (1970) (1) Time translations in the algebraic formulation of statistical mechanics. *J. math. Phys.* **11.**

and STREATER, R. F. (1967) (1) Nonexistence of the ferromagnetic continuum. *Nuovo Cim.* Ser. X **50.**

EINSTEIN, A. (1924, 1925) (1) Quantentheorie des einatomigen Gases. *Berl. Ber. Akad.* Sept. 20 issue, p.261 (1924), Feb. 9 issue, p.3 (1925).

EMCH, G. G. and GUENIN, M. (1966) (1) Gauge invariant formulation of the BCS-model. *J. math. Phys.* **7.**

and KNOPS, H. J. F. (1970) (1) Pure thermodynamical phases as extremal KMS states. *J. math. Phys.* **11.**

and VERBOVEN, E. J. (1968) (1) On the extension of invariant partial states in statistical mechanics. *Commun. Math. Phys.* **7.** (1968) (2) On partial weakly clustering states with applications to the Ising model. *Commun. Math. Phys.* **8.** (1970) (3) The breaking of euclidean symmetry with an application to the theory of crystallization. *J. math. Phys.* **11.**

and RADIN, C. (1971) (1) Relaxation of local thermal deviations from equilibrium. *J. math. Phys.* **12.**

FERMI, E. (1926) (1) Zur Quantelung des idealen einatomigen Gases. *Z. Phys.* **36**; Sulla quantizzazione del gas perfette monoatomico. *Atti Accad. naz. Lincei Rc.* **3.**

FOCK, V. (1932) (1) Konfigurationsraum und zweite Quantelung. *Z. Phys.* **75.**

FRÖHLICH, H. (1950) (1) Theory of the superconducting state I. *Phys. Rev.*, **79.** (1950) (2) Theory of the superconducting state II. *Proc. phys. Soc.* A **63.**

GEL'FAND, I. M. and NAIMARK, M. A. (1943) (1) On the imbedding of normed

rings into the ring of operators in Hilbert space. *Mat. Sb.* N.S. **12**, [54], 197.
GILLE, J. F. and MANUCEAU, J. (1972) (1) Guage transformations of second type and their implementations–I Fermions. *J. math. Phys.* **13**.
GINIBRE, J., GROSSMANN, A., and RUELLE, D. (1966) (1) Condensation of lattice gases. *Commun. math. Phys.* **3**.
GUERRA, F., ROSEN, L., and SIMON, B. (1972) (1) Nelson's symmetry and the infinite volume behaviour of the vacuum in $P(\Phi)_2$. *Commun. math. Phys.* **27**.
GUICHARDET, A. (1966) (1) Produits tensoriels infinis et représentations des relations d'anticommutation. *Annls. scient. Éc. norm. sup., Paris* **83**.
HAAG, R. (1962) (1) The mathematical structure of the BCS-model. *Nuovo Cim.* **25**.

HUGENHOTZ, N. M., and WINNINK, M. (1967) (1) On the equilibrium states in quantum statistical mechanics. *Commun. math. Phys.* **5**.

and KASTLER, D. (1964) (1) An algebraic approach to quantum field theory. *J. math. Phys.* **5**.

and SCHROER, B. (1962) (1) Postulates of quantum field theory. *J. math. Phys.* **3**.

HEISENBERG, W. (1926) (1) Schwankungserscheinungen und Quanten Mechanik. *Z. Phys.* **40**. (1928) (2) Zur Theorie des Ferromagnetismus. *Z. Phys.* **49**.
HEPP, K. and LIEB, E. H. (1973) (1) On the super radiant phase transition for molecules in a quantized radiation field; the Dicke maser model. *Ann. Phys.* **76**. (1973) (2) The equilibrium statistical mechanics of matter interacting with the quantized radiation field. Preprint: Sem. für theor. Physik, Eidgenössische Technische Hochschule, Zürich. (Phys./73/049).
HERMAN, R. H. and TAKESAKI, M. (1970) (1) States and automorphism groups of operator algebras. *Commun. math. Phys.* **19**.
HUGENHOLTZ, N. M. (1967) (1) On the factor type of equilibrium states in quantum statistical mechanics. *Commun. Math. Phys.* **6**.

and WIERINGA, J. D. (1969) (1) On locally normal states in quantum statistical mechanics. *Commun. math. Phys.* **11**.

INÖNÜ, E. and WIGNER, E. P. (1953) (1) On the contraction of groups and their representations. *Proc. natn. Acad. Sci.* U.S.A. **39**.
JELINEK, F. (1968) (1) BCS-Spin-Model, its thermodynamic representations and automorphisms. *Commun. math. Phys.* **9**.
JORDAN, P. and WIGNER, E. P. (1928) (1) Über das Paulische Aquivalenzverbot. *Z. Phys.* **47**.
KAMERLINGH-ONNES, H. (1911) (1) *Leiden Comm.* 122b, 124c; Suppl. 35, **193**.
KASTLER, D. (1965) (1) The C*-algebras of a free boson field–I Discussion of the basic facts. *Commun. math. Phys.* **1**.

HAAG, R. and MICHEL, L. (1967/1968) (1) *Central decomposition of ergodic states.* Séminaire de Physique théoretique, I. Marseille.

MEBKHOUT, M., LOUPIAS, G., and MICHEL, L. (1972) (1) Central decomposition of invariant states. Applications to the groups of time translations and of euclidean transformations in algebraic field theory. *Commun. math. Phys.* **27**.

POOL, J. C. T., and THUE POULSEN, E. (1969) (1) Quasi-unitary algebras attached to temperature states in statistical mechanics. A comment on the work of Haag, Hugenholtz and Winnink. *Commun. math. Phys.* **12.**

and ROBINSON, D. W. (1966) (1) Invariant states in statistical mechanics. *Commun. math. Phys.* **3.**

KAUFMAN, B. (1949) (1) Crystal statistics II. Partition function evaluated by spinor analysis. *Phys. Rev.* **76.**

KLAUDER, J. R. and STREIT, L. (1969) (1) Properties of 'quadratic' canonical commutation relation representations. *J. math. Phys.* **10.**

KUBO, R. (1957) (1) Statistical mechanical theory of irreversible processes, I. *J. phys. Soc. Japan,* **12.**

LANFORD, O. E. and ROBINSON, D. W. (1968) (1) Statistical mechanics of quantum spin systems III. *Commun. math. Phys.* **9.** (1968) (2) Mean entropy of states in quantum statistical mechanics. *J. math. Phys.* **9.**

and RUELLE, D. (1967) (1) Integral representations of invariant states on B*-algebras. *J. math. Phys.* **8.**

LEWIS, J. T. (1972) (1) The free boson gas. *Mathematics of contemporary physics* Ed. R. F. Streater Academic Press, London and New York.

and PULE, J. V. (1972) (1) *The equilibrium states of the free Bose gas.* Preprint: Oxford University.

LIEB, E. H. (1973) (1) The classical limit of quantum spin states. *Commun. math. Phys.* **1.**

LIMA, R. and VERBEURE, A. (1971) (1) Local perturbations and approach to equilibrium. Preprint: 71, p.414 Centre de physique théoretique, Marseilles. (1971) (2) Bilinear Hamiltonians and ergodicity. The linear *XY*-Model. Preprint: Centre de physique théoretique, Marseilles.

MANUCEAU, J. (1968) (1) C*-algébre de relations du commutation. *Annls. Inst. Henri Poincaré* **8.**

ROCCA, F. and TESTARD, D. (1969) (1) On the product form of quasi-free states. *Commun. math. Phys.* **12.**

SIRUGUE, M., TESTARD, D., and VERBEURE, A. (1968) (1) The smallest C*-algebra for canonical commutation relations. Special lecture: 1971 Europhysics Conference, Haifa. Cf. Manuceau (1).

and VERBEURE, A. (1968) (1) Quasi-free states of the CCR-algebra and Bogoliubov transformations. *Commun. math. Phys.* **8.** (1970) (2) Non-factor quasi-free states of the CAR-algebra. *Commun. math. Phys.* **18.**

MARINARO, M. and SEWELL, G. L. (1972) (1) Characterizations of phase transitions in Ising spin systems. *Commun. math. Phys.* **24.**

MARTIN, P. C. and SCHWINGER, J. (1959) (1) Theory of many-particle systems I. *Phys. Rev.* **115.**

MARTIN-LÖF, A. (1973) (1) Mixing properties, differentiability of the free energy and the central limit theorem for a pure phase in the Ising model at low temperatures. *Commun. math. Phys.* **32.**

MEISSNER, W. and OCHSANFELD, R. (1933) (1) Ein neuer Effekt bei eintritt der Supraleifähigkeit. *Naturwissenschaften* **21**

MOYA, R. P. (1973) (1). *Equilibrium states for the infinite free Bose gas.* Preprint: Queen Mary College, London (1973); Thesis (1973).

NELSON, E. (1973) (1) Construction of quantum fields from Markoff fields.

J. Funct. Anal. **12.** (2) The free Markoff field. *J. funct. Anal.* **12.**

ONSAGER, L. (1944) (1) Crystal statistics I. A two-dimensional model with an order-disorder transition. *Phys. Rev.* **65.**

OSTERWALDER, K. and SCHRADER, R. (1973) (1) Axioms for Euclidean Green's functions. *Commun. math. Phys.* **31.**

PAULI, W. (1925) (1) Über den Zusammenhang des Abchlusses der Elektronengruppen im Atom mit der komplex Struktur der Spektren. *Z. Phys.* **31.** (1927) (2) Über Gasentartung und Paramagnetismus. *Z. Phys.* **41.** (1927) (3) Zur Quantenmechanik des magnetischen Elektrons. *Z. Phys.* **43.** (1940) (4) The connection betwwen spin and statistics. *Phys. Rev.* **58.**

POWERS, R. T. (1967) (1) Representations of uniformly hyperfinite algebras and their associated Von Neumann rings. *Bull. Am. math. Soc.* **73**; *Ann. Math.* **86.** (1967) (2) *Representations of the canonical anticommutation relations.* Thesis: Princeton University.

and STØRMER, E. (1970) (1) Free states of the canonical anticommutation relations. *Commun. math. Phys.* **16.**

PRESUTTI, E., SCACCIATELLI, E., SEWELL, G. L. and WANDERLINGH, F. (1972) (1) Studies in the C*-algebraic theory of non-equilibrium statistical mechanics: Dynamics of open and of mechanically-driven systems. *J. math. Phys.* **13.**

RADIN, C. (1970) (1). Approach to equilibrium in a simple model. *J. math. Phys.* **11.** (1971) (2) Noncommutative mean ergodic theory. *Commun. math. Phys.* **21.** (1971) (3) Gentle perturbations. *Commun. math. Phys.* **23.** (1972) (4) Automorphisms of Von Neumann algebras as point transformations. Preprint: Joseph Henry Laboratories of Physics, Princeton University. To appear in *Proc. Am. Math. Soc.* (1972) (5) Ergodicity in Von Neumann algebras. Preprint: Princeton University. To appear in *Pacif. J. Math.* (1973) (6) Dynamics of limit models. *Commun. math. Phys.* **33.**

ROBINSON, D. W. (1965) (1) The ground state of the Bose gas. *Commun. Math. Phys.* **1.** (1967) (2) Statistical mechanics of quantum spin systems I. *Commun. math. Phys.* **6.** (1967). (1968) (3) Statistical mechanics of quantum spin systems II. *Commun. math. Phys.* **7.** (1968). (1969) (4) A proof of the existence of phase transitions in the anisotropic Heisenberg model. *Commun. math. Phys.* **14.** (1970) (5) Normal and locally normal states. *Commun. math. Phys.* **19.**

and RUELLE, D. (1967) (1) Extremal invariant states. *Annls Inst. Henri Poincaré.* **6.**

ROCCA, F. and SIRUGUE, M. (1973) (1) *Phase operators and condensed systems.* Preprint: (N Th 73/4) Laboratoire de Physique Théoretique, Université de Nice.

and TESTARD, D. (1969) (1) On a class of equilibrium states under the Kubo-Martin-Schwinger boundary condition. I. Fermions. *Commun. math. Phys.,* **13.** (1970) (2) II. Bosons. *Commun. math. Phys.* **19.**

RUELLE, D. (1966) (1) States of physical systems. *Commun. math. Phys.,* **3.** (1967) (2) States of classical statistical mechanics. *J. math. Phys.* **8.** (1970) (3) Integral representation of states on a C*-algebra. *J. funct. Anal.* **6.**

RUSKAI, M. -B. (1971) (1) Time evolution of quantum lattice systems. *Commun. math. Phys.* **20.**

SCHAFROTH, M. R. (1960) (1) Theory of superconductivity. *Solid St. Phys.* **10.**

SCHULTZ, T. D., MATTIS, D. C., and LIEB, E. H. (1964) (1) Two-dimensional Ising model as a soluble problem of many fermions. *Rev. mod. Phys.* **36.**

SCHWINGER, J. (1958) (1) On the Euclidean structure of relativistic field theory. *Proc. natn. Acad. Sci. U.S.A.* **44.**

SEGAL, I. E. (1947) (1) Postulates for general quantum mechanics. *Annls Math.* **48.** (1951) (2) A class of operator algebras which are determined by groups. *Duke math. J.* **18.** (1953) (3) A noncommutative extension of abstract integration. *Annls Math.* **57, 58.** (1959, 1961, 1962) (4) Foundations of the theory of dynamical systems of infinitely many degrees of freedom I. *Math-fys. Meddr.* **31** (1959). II. *Can. J. Math.* **13** (1961). III. *Illinois J. Math.* **6** (1962).

SEWELL, G. L. (1973) (1) Ergodic theory in algebraic statistical mechanics. In *Lectures in theoretical physics*, Vol. 12. Eds. K. T. Mahanthappa and W. E. Brittin. Gordon and Breach. (1973) (2) States and dynamics of infinitely extended physical systems. *Commun. math. Phys.* **33.**

SIRUGUE, M. and TESTARD, D. (1971) (1) Some connections between ground states and temperature states of thermodynamical systems. *Commun. math. Phys.,* **22.**

—— and WINNINK, M. (1970) (1) Constraints imposed upon a state of a system that satisfies the KMS boundary condition. *Commun. math. Phys.* **19.** (1971) (2) Translations dans le temps comme group D'*-automorphismes. Preprint: No. 71, p.388. Centre de Physique Théoretique, Marseille.

SLAWNY, J. (1972) (1) On the von Neumann uniqueness theorem and C*-algebras for quantum systems. Preprint: Tel-Aviv University. To appear in *Trans. Am. math. Soc.*

STREATER, R. F. (1967) (1) The Heisenberg ferromagnet as a quantum field theory. *Commun. math. Phys.* **6.**

SYMANZIK, K. (1969) (1) Euclidean quantum field theory. *Proceedings of the International School of Physics 'Enrico Fermi', Varenna Course* XLV (Ed. R. Jost). Academic Press, New York.

TAKEDA, Z. (1955) (1) Inductive limit and infinite direct product of operator algebras. *Tohoku math. J.* **7.**

TAKESAKI, M. (1970) (1) Disjointness of the KMS-states of different temperatures. *Commun. math. Phys.* **17.**

—— and WINNINK, M. (1973) (1) Local normality in quantum statistical mechanics. *Commun. math. Phys.* **30.**

THIRRING, W. (1968) (1) On the mathematical structure of the BCS-model. II: *Commun. math. Phys.* **7.**

—— and WEHRL, A. (1967) (1) On the mathematical structure of the BCS-model, I: *Commun. math. Phys.* **4.**

TISZA, L. (1938) (1) Transport phenomena in Helium II. *Nature Supplement* May 21.

UHLENBECK, G. E. and GOUDSMIT, S. (1925) (1). Ersetzung der Hypothese vom unmechanischen zwang durch eine Forderung bezüglich des inneren Verhaltens jedes einzelnen Elektrons. *Naturwissenschaften* **13.** (1926) (2) Spinning electrons and the structure of spectra. *Nature* **117.**

VAN DONGEN, J. M. M. and VERBOVEN, E. J. (1973) (1) *Decomposition of KMS states*. Preprint: Institute of theoretical Physics, University of Nijmegan.

VERBEURE, A. and VERBOVEN, E. J. (1966) (1) Quantum states of an infinite system of harmonic oscillators with linear interaction terms. *Phys. Letters* **23.** (1967) (2) States of infinitely many oscillators. *Physica* **37.** (1967) (3) States of infinitely many oscillators with linear interaction terms. *Physica* **37.**

VERBOVEN, E. J. (1966) (1,2) Quantum thermodynamics of an infinite system of harmonic oscillators–I. *Phys. Letters* **21**; –II. *Physica* **32.**

WIERINGA, J. D. (1970) (1). *Thermodynamic limit and KMS states in quantum statistical mechanics. A C*-algebraic approach.* Thesis: Groningen University.

WINNINK, M. (1969) (1) Algebraic aspects of the Kubo-Martin-Schwinger condition. *Lectures: Cargèse Summer School.*

Texts and Monographs

ABRIKOSOV, A. A., GOR'KOV, L. P., and DZYALOSHINSKII, I. YE. (1965) [1] *Quantum field theoretical methods in statistical physics.* Pergamon Press, Oxford and London.

ARNOLD, V. I. and AVEZ, A. (1968) [1] *Ergodic problems of classical mechanics.* Benjamin, New York.

AUSLANDER, L. (1967) [1] *Differential geometry.* Harper and Row, London.

BOGOLIUBOV, N. N., TOLMACHEV, V. V., and SHIRKOV, D. V. (1958) [1] *Izv. Akad. Nauk SSSR.* Translated as *A new method in the theory of superconductivity.* Consultants Bureau; Plenum Press, New York.

BOLTZMANN, L. (1968) [1] *Abhandlungen.* Reprinted: Chelsea Publishing Company, New York (originally published in 1909). (1964) [2] *Vorlesungen über Gastheorie.* Reprinted: University of California Radiation Laboratories, Berkeley, and Cambridge University Press, London (originally published in 1910).

BORN, M. (1957) [1] *Atomic physics.* Hafner Publishing Company, New York.

BRUSH, S. G. (1964) [1] *History of the Lenz-Ising model.* University of California Radiation Laboratories, Report 7940.

CHOQUET, G. (1969) [1] *Lectures on analysis* (Ed. J. Marsden, T. Lance, and S. Gelbart) Vols I-III. Benjamin, New York.

DIXMIER, J. (1964) [1] *Les C*-algèbres et leurs représentations.* Gauthier-Villars, Paris. (1969) [2] *Les algèbres d'opérateurs l'espace hilbertien* (2e edition). Gauthier-Villars, Paris.

ECKMANN, J.-P. and GUENIN, M. (1969) [1] *Méthodes algébriques en mécanique statistique.* In lecture notes in mathematics No. 81. Springer-Verlag, Berlin, Heidelberg, and New York.

EHRENFEST, P. and EHRENFEST, T. (1959) [1] *The conceptual foundations of the statistical approach in mechanics.* Reprinted: Cornell University Press, Ithaca (originally published in 1901).

EMCH, G. G. (1972) [1] *Algebraic methods in statistical mechanics and quantum field theory.* Wiley-Interscience, New York.

GAMOW, G. (1967) [1] *One two three . . . infinity.* Bantam Books Inc., New York.

GANTMACHER, F. R. (1960) [1] *Matrix theory* Vols I and II. Chelsea Publishing Company, New York.

GEL'FAND, I. M.; RAIKOV, D. A.; and SHILOV, G. E. (1964) [1] *Commutative normed rings.* Chelsea Publishing Company, New York.

—— and SHILOV, G. E. (1964, 1968) [1,2] *Generalized functions* Vols 1 and 2. Academic Press, New York and London.

and VILENKIN, N. Y. (1964) [1] *Generalized functions* Vol. 4. Academic Press, New York and London.

GIBBS, J. W. (1960) [1] *Elementary principles of statistical mechanics.* Reprinted: Dover, New York (originally published in 1902).

GREENLEAF, F. P. (1969) [1] *Invariant means on topological groups.* Van Nostrand Reinhold Company, New York.

GROTHENDIECK, A. (1955) [1] Produits tensoriels topologiques et espaces nucleaires. *Mem. Am. math. Soc.* No. 16. American Mathematical Society Providence, Rhode Island.

GUGGENHEIM, E. A. (1955) [1] *Boltzmann's distribution law.* North Holland Publishing Company, Amsterdam.

GUICHARDET, A. (1969) [1,2] *Tensor products of C*-algebras.* Aarhus Matematische Institut, Lecture Universität Series Notes. Nos. 12, 13. (1972) [3] Symmetric Hilbert spaces. In *Lecture notes in mathematics* Vol. 261. Springer-Verlag, Berlin, Heidelberg, and New York. (1968) [4] *Algèbras d'observables associées aux relations de commutation.* A. Colin, Paris.

HELMBERG, G. (1969) [1] *Introduction to spectral theory in Hilbert space.* North-Holland Publishing Company, Amsterdam.

HILLE, E. and PHILLIPS, R. S. (1954) [1] *Functional analysis and semi-groups.* American Mathematical Society Providence.

HUANG, K. (1963) [1] *Statistical mechanics.* Wiley and Sons, New York and London.

KATO, T. (1966) [1] *Perturbation theory for linear operators.* Springer-Verlag; Berlin, Heidelberg, and New York.

LANDAU, L. D. and LIFSCHITZ, E. M. (1958) [1] *Statistical physics.* Pergamon Press, Oxford and London.

LONDON, F. (1961) [1] Superfluids. *Macroscopic theory of superconductivity* Vol I. Dover, New York.

MATTIS, D. C. (1965) [1] *The theory of magnetism.* Harper and Row, London.

NARASIMHAN, R. (1971) [1] Several complex variables. In *Chicago lectures in mathematics.* The University of Chicago Press.

PARTHASARATHY, K. R. and SCHMIDT, K. (1972) [1] Positive definite kernels, continuous tensor products, and central limit theorems of probability theory. In *Lecture notes in mathematics* Vol 272. Springer-Verlag, Berlin, Heidelberg, and New York..

PHELPS, R. R. (1966) [1] *Lectures on Choquet's theorem.* Van Nostrand Reinhold Company, New York.

RIESZ, F. and SZ-NAGY, B. (1960) [1] *Functional analysis.* Frederick Unger Publishing Company, New York.

ROBERTSON, A. P. and ROBERTSON, W. J. (1964) [1] *Topological vector spaces* Cambridge University Press, London.

RUELLE, D. (1969) [1] *Statistical mechanics, rigorous results.* Benjamin, New York.

SAKAI, S. (1971) [1] *C*-algebras and W*-algebras.* Springer-Verlag, Berlin, Heidelberg, and New York.

SCHRÖDINGER, E. (1957) [1] *Statistical thermodynamics.* Cambridge University Press, London.

SEGAL, I. E. (1963) [1] *Mathematical problems of relativistic physics.* American Mathematical Society, Providence.

and KUNZE, R. (1969) [1] *Integrals and operators.* McGraw-Hill, New York.

SPANIER, E. H. (1966) [1] *Algebraic topology.* McGraw-Hill, New York.

STERNBERG, S. (1964) [1] *Lectures on differential geometry.* Prentice-Hall, Englewood Cliffs, New Jersey.

STREATER, R. F. and WIGHTMAN, A. S. (1964) [1] *PCT, spin and statistics, and all that.* Benjamin, New York.

TAKESAKI, M. (1970) [1] Tomita's theory of modular Hilbert algebras and its applications. In *Lecture notes in mathematics* No. 128. Springer-Verlag, Berlin, Heidelberg, and New York.

TRÈVES, F. (1967) [1] *Topological vector spaces, distributions, and kernels.* Academic Press, New York.

VARADARAJAN, V. S. (1970) [1] *Geometry of quantum theory* Vol II. Van Nostrand Reinhold Company, New York.

VILENKIN, N. YA. (1968) [1] Special functions and the theory of group representations. In *Translations of mathematical monographs* Vol 22. American Mathematical Society, Providence.

VAN VLECK, J. H. (1932) [1] *The theory of electric and magnetic susceptibilities.* Clarendon Press, Oxford.

Author Index

Abraham, D. B. 91
Abrikosov, A. A. 50
Araki, H. 3, 15, 21, 39, 50, 91
Arnold, V. I. 8
Auslander, L. 3, 22, 30
Avez, A. 8

Bardeen, J. 70, 79
Baumann, K. 70
Bogoliubov, N. N. 70, 80
Boltzmann, L. iii, 1
Born, M. iii, 20, 89
Bose, S. N. 39, 49
Brascamp, H. J. 91
Brush, S. G. 90

Cannon, J. T. 39, 40, 41, 45, 50, 60, 63
Cantor, G. 31
Chaiken, J. M. 5, 26
Choquet, G. 3, 6, 7, 10, 11, 12, 21, 22, 27, 35, 52, 56, 67, 101
Cook, J. M. 5, 22, 28, 34
Cooper, L. N. 70, 71, 79

Davies, E. B. 2, 10, 82
Dell'Antonio, G.-F. 3, 6, 26, 44, 49, 55, 69
Dirac, P. A. M. 38, 39, 48, 89
Dixmier, J. 3, 4, 6, 7, 8, 14, 18
Doplicher, S. 6, 8, 26, 32
Dubin, D. A. 3, 13, 14, 17, 91, 97
Dzyaloshinskii, I. Ye. 50

Eckmann, J.-P. iii, 3
Eder, G. 70
Ehrenfest, P. iii, 1, 20
Ehrenfest, T. iii, 1, 20
Einstein, A. 39, 49
Emch, G. G. iii, 2, 3, 4, 6, 7, 8, 9, 12, 13, 15, 18, 22, 24, 29, 34, 80, 94, 96, 99, 106

Fermi, E. 20, 38, 39, 71
Fock, V. 5, 22, 28
Fröhlich, H. 70

Gallavotti, G. 91
Gamow, G. 31
Gantmacher, F. R. 104
Gel'fand, I. M. 3, 7, 10, 15, 22, 27, 35, 56, 67, 101
Gibbs, J. W. iii, 1
Gille, J. F. 26, 49, 55
Ginibre, J. 91
Gor'kov, L. P. 50
Goudsmidt, S. 89
Greenleaf, F. P. 9
Grossmann, A. 91
Grothendieck, A. 3, 30
Guenin, M. iii, 3, 80
Guerra, F. 100
Guggenheim, E. A. iii
Guichardet, A. 3, 4, 22, 23, 24, 29, 30

Haag, R. 3, 4, 8, 13, 14, 16, 80
Helmberg, G. 13, 15
Heisenberg, W. 89, 90
Hepp, K. 82
Herman, R. H. 8
Hille, E. 14, 15, 68
Huang, K. iii, 20, 90, 100, 103
Hugenholtz, N. M. 3, 6, 8, 13, 14, 15, 16

Inönü, E. 78

Jelinek, F. 78, 85, 86
Johnson, S. iii
Jordan, P. 29

Kac, M. 39, 41, 45, 57, 61
Kadison, R. V. 8
Kamerlingh-Onnes, H. 70
Kastler, D. 3, 4, 8, 14, 16
Kato, T. 12, 14, 15, 26, 33, 34
Kaufman, B. 100
Klauder, J. R. 50
Knops, H. J. F. 15, 106
Kubo, R. 14
Kunze, R. 3

Landau, L. D. iii, 20, 50, 62, 71
Lanford, O. E. 8, 9, 67, 68, 91
Lewis, J. T. 21, 39, 40, 50, 69
Lieb, E. 82, 90, 100
Lifschitz, E. M. iii, 20, 50, 62, 71
Lima, R. 2

London, F. iii, 49, 71
Loupias, G. 8

Manuceau, J. 4, 21, 26, 28, 49, 55
Marinaro, M. 90, 100, 103, 104, 105
Martin, P. C. 14
Martin-Löf, A. 91
Mattis, D. C. iii, 89, 90, 100
Mebkhout, M. 8
Meissner, W. 70
Michel, L. 8
Miyata, H. 15
Moya, R. P. 2, 6, 19, 21, 69

Naimark, M. A. 7
Narasimhan, R. 77
Nelson, E. 100

Ochsanfeld, R. 70
Onsager, L. 90, 98
Osterwalder, K. 100

Parthasarathy, K.R. 56
Pauli, W. 20, 31, 48, 89
Peierls, R. 99
Phelps, R. R. 6, 10
Phillips, R. S. 14, 15, 68
Pool, J. C. T. 14, 16
Powers, R. T. 21, 28, 32
Presutti, E. 2
Pulé, J. V. 21, 39, 40, 50, 69

Radin, C. 2, 8, 98
Raikov, D. A. 101
Riesz, F. 9, 14, 22
Robertson, A. P. 3, 11, 35, 52, 67
Robertson, W. J. 3, 11, 35, 52, 67
Robinson, D. W. 3, 5, 8, 50, 67, 68, 90, 91, 94, 96, 99
Rocca, F. 15, 21, 26, 28, 29, 49
Rosen, L. 100
Ruelle, D. iii, 3, 4, 6, 7, 8, 9, 12, 13, 22, 24, 26, 91, 96
Ruskai, M.-B. 13, 96

Sakai, S. 3, 4, 5, 6, 7, 8, 9, 13, 14, 18, 21, 30, 32, 85, 94
Scacciatelli, E. 2
Schafroth, M. R. 71
Schmidt, K. 56
Schrader, R. 100
Schrieffer, J. R. 70, 79
Schrödinger, E. iii, 20, 35, 48
Schroer, B. 3, 4
Schultz, T. D. 90, 100
Schwartz, L. 15
Schwinger, J. 14, 100
Segal, I. E. 3, 4, 7, 8, 52
Sewell, G. L. 2, 3, 8, 13, 14, 17, 19, 21, 69, 90, 97, 100, 103, 104, 105
Sexl, R. 70
Shilov, G. E. 15, 27, 35, 67, 101
Shirkov, D. V. 70
Simon, B. 100
Sirugue, M. 4, 13, 15, 21, 26, 28, 29, 49
Slawny, J. 4
Spanier, E. H. 11
Sternberg, S. 3, 22, 30
Stone, M. H. 12, 68
Størmer, E. 8, 21, 29
Streater, R. F. 15, 24, 48, 53, 90, 91
Streit, L. 50
Symanzik, K. 100
Sz-Nagy, B. 9, 14, 22

Takeda, Z. 4
Takesaki, M. 8, 14, 18, 47, 69, 87, 97
Testard, D. 4, 15, 21, 29
Thirring, W. 70, 78, 86, 87
Thue Poulsen, E. 14, 16
Tisza, L. 49

Tokmachev, V. V. 70
Tomita, M. 47, 87
Trêves, F. 15, 27, 35, 52, 67

Uhlenbeck, G. E. 89

Van Dongen, J. M. M. 8, 10
Van Vleck, J. H. 90
Varadarajan, V. S. 6, 53
Verbeure, A. 2, 4, 21, 28, 50
Verboven, E. J. 8, 10, 15, 50, 106
Vilenkjn, N. Ya. 3, 10, 22, 27, 56, 65, 74, 75, 78

Wanderlingh, F. 2
Wehrl, A. 87
Weyl, H. 52
Wieringa, J. D. 4, 6, 8, 13, 53
Wightman, A. S. 15, 24, 48, 53, 97
Wigner, E. P. 29, 78
Wils, W. 10
Winnink, M. 3, 13, 14, 15, 16, 18, 97
Woods, E. J. 3, 50
Wyss, W. 3, 21, 39

Subject Index

abelian algebra, *see* algebra
  asymptotic, q.v.
absorbing family, 3, 21, 50, 91, 100
algebra
  abelian, 101–102
  automorphisms, q.v.
  C^*-, q.v.
  decomposition of, *see*
  global decomposition
  direct integral of, 64, 66, 84–85
  local, 4, 27, 93
  local subalgebras, 3
  also see BCS model, CAR, CCR,
  lattice models
  modular Hilbert, q.v.
  quasilocal, q.v.
  simple, 32
  spin, *see* CAR
  tensor products, q.v.
  uniformly hyperfinite (UHF), q.v.
  Von Neumann, *see* W^*-algebra
  W^*-, q.v.
  Weyl, q.v.
annihilation operators, see CAR, CCR
asymptotic abelian, 12
  G-, 9
automorphism
  gauge group of, q.v.
  group of, q.v.
  inner, 14
  space translations, q.v.
  time translations, q.v.
  unitary implementability, 8, 12, 13, 17

BCS (Bardeen, Cooper, Schrieffer)
model, 70 ff.
  Hilbert spaces, q.v.
  local subalgebras, 72
  quasilocal algebra, 72
bicommutant of an algebra, 13
Bose-Einstein condensation, 39, 66
Bose gas, *see* ideal Bose gas
boson, 48
  fields, 51
  Fock-Cook space, q.v.
boundary condition, toroidal, 31, 102

C^*-algebra, 3
  for the CAR, q.v.
  for the CCR, q.v.
  of quasilocal observables, q.v.
C^*-inductive limit, q.v.
canonical ensemble state, 44–45, 49, 63
CAR (canonical anticommutation
relations), 24
  even algebras, 25
  Fock-Cook representation, 24, 29
  local subalgebras, 24–25
  quasilocal algebra, 27
  spin representation, *also see* spin, 29
CCR (canonical commutation relations), 52
  Fock-Cook representation, 52
  local subalgebras, 53–54
  quasilocal algebra, 54
  Weyl form, 52
central decomposition, 10, 69
commutant of an algebra, 13
critical density, 62
cubical local regions, 21
cyclic
  representation, q.v.
  vector, 8

density
  Bose-Einstein operator, 59
  Fermi-Dirac operator, 38, 42
  matrix, *also see* normal state of
  particles, 5
  also see equation of state
dynamics
  see time translations

electrons, *see* fermions
equation of state, 41, 61
ergodic
  decomposition, 10, 12, 63, 106
  KMS decomposition, 69
  states, q.v.
    theorem, 9
Euler angles, 74

factor, *see* W^*-algebras
Fermi gas, *see* ideal Fermi gas
fermion, 20

fields, 24, 27
Fock-Cook space, q.v.
Fock-Cook
condition, 28, 55
space, 5
antisymmetric, 22
even subspace, 23
symmetric, 50
state, 28, 55
vacuum vector, 28, 55
fugacity, 40, 57, 60

G-abelian property, 9
gap equation, 78
gauge group
for the CAR, 25
for the CCR, 55, 66
and fields, 26
and the number operator, 25, 39, 49, 55
spontaneous symmetry breakdown, 64, 88

Gibbs state
global, 16, 39, 49, 59, 76
local, 15, 36, 57, 74, 102, 104
global decomposition, *see* ergodic decomposition
GNS (Gel'fand and Naimark; Segal)
construction, 7
representation, 7, 17, 43, 63, 84
grand canonical ensemble, 14
partition function, 15, 37
state, *see* Gibbs state
group
of automorphisms, 6
contraction, 78, 81
gauge, q.v.
invariance, *see* state
projective representations, 53

Hamiltonian, 14, 95–96
BCS, 72, 76
Haag-Bogoliubov, 80
Heisenberg, 97
Ising, 98, 102
one particle global, 34
one particle local, 32
reduced, 13, 33, 35
Heisenberg model, 90
helium isotopes, 49
Hilbert space, 3, 22
direct integrals, 10, 64, 66, 84–85
direct sums, 22, 51, 75
inductive limits, q.v.
local subspaces, 3, 22, 50, 71, 92
tensor products, 6, 43, 71, 83, 92

ideal
Bose gas, 48 ff.
Fermi gas, 20 ff.
incomplete direct product space, *see* Hilbert space tensor products
inductive limits
for C^*-algebras, 4, 27, 54, 72, 93
for Hilbert spaces, 3, 23, 51, 71, 81, 92
injective mappings, 11, 22, 27, 51, 53, 71, 72, 82–83, 92–93
Ising model, 90, 99 ff.
generalized, 98
isotony
for algebras, 4, 27, 54, 72
for Hilbert spaces, 3, 23, 51, 71

Kac density, 40, 57, 61
KMS condition, 14, 46, 69

lattice, 91, 100
models, 89
Hilbert spaces, q.v.
local subalgebras, 92, 100–102
quasilocal algebras, 93, 100–103
Liouville equation, 46, 68
local
algebras, q.v.
commutativity, 6, 25, 27, 54
Hilbert spaces, q.v.
regions, 3, 21
subalgebras, *see* algebras

mean field models, 81–82
modular Hilbert algebras, 47, 69, 87

nearest neighbours, 98
number operator, *see* gauge group

observables
localizable, 4
systemic, 4

Pauli
exclusion principle, 20, 48
matrices, 31, 72
periodic boundary conditions, q.v.
phonons, 49
powers' factor, 42

quasilocal algebra, *also see* BCS model, CAR, CCR, lattice model, 4
quasifree evolution, *see* Hamiltonian, one particle
state, q.v.

representation
cyclic, 7–8
equivalence, 2, 8
also see thermodynamic representations
row algebras, 101

second sound, 49
space translations, 10
for CAR algebras, 26–27
for CCR algebras, 54–55
for lattice models, 93–95
spin
algebra, *also see* CAR, 48, 72, 89
operators, 31, 72, 99, 102, 103
reversal symmetry, 105
states, 4
even, 29
ergodic, 7–8
extremal, 6
extremal KMS, 18
factorial, 94
G-extremal, 7, 8
G-invariant, 7
locally normal, 5, 8
normal, 5
primary, 94
product, 44, 94, 105
quasifree, 28, 38, 56
vector, 5
SU(2) group, 74
representations, 74
superfluids, *see* two-fluid model
symmetry, 6
group of, q.v.
invariance of states, *see* states
spontaneous breakdown of, 10
symplectic form, 52–53

tensor product
of C^*-algebras, 30, 43, 83, 92
of Hilbert spaces, q.v.
thermodynamic representations
BCS model, 82
ideal Bose gas, 63
ideal Fermi gas, 43
time translations
convergance of, 17, 35, 86, 96
global, 17, 34, 67, 85
local, 13, 32, 56, 94
toroidal boundary conditions, q.v.
transfer matrix, 103–104
two-fluid model, 49

UHF (uniformly hyperfinite) algebras, 94

W^*-algebra, 3, 13, 17, 42, 47, 67–69, 85, 87, 94
also see thermodynamic representations
Weyl fields, 52
also see CCR

XY-model, 90, 98

The study of exactly solvable models contributes to the understanding of theoretical structures devised to explain physical phenomena. The structure examined in this book is quantum statistical mechanics, adapted to systems with an infinite number of degrees of freedom. The standard solvable models of statistical mechanics are analysed in depth : the ideal Fermi gas, the ideal Bose gas, the BCS model of superconductivity, and the Ising model.

This book will be of interest to physics postgraduate students beginning research projects in this field, to research workers in statistical mechanics, and to functional analysts who are interested in applications of their subject.

In spite of the mathematical nature of the material, very few proofs are given. It is hoped that this book will enable the reader who needs proofs to find his way through the journals and the reader who does not wish to see proofs to understand some of the technical complications of the subject.

OXFORD UNIVERSITY PRESS

ISBN 0 19 853341 1